LECTURES ON RANDOM LOZENGE TILINGS

Over the past 25 years, there has been an explosion of interest in the area of random tilings. The first book devoted to the topic, this timely text describes the mathematical theory of tilings. It starts from the most basic questions (which planar domains are tileable?) before discussing advanced topics about the local structure of very large random tessellations. The author explains each feature of random tilings of large domains, discussing several different points of view and leading on to open problems in the field. The book is based on upper-division courses taught to a variety of students, but it also serves as a self-contained introduction to the subject.

Test your understanding with the exercises provided, and discover connections to a wide variety of research areas in mathematics, theoretical physics, and computer science, such as conformal invariance, determinantal point processes, Gibbs measures, high-dimensional random sampling, symmetric functions, and variational problems.

Vadim Gorin is a faculty member at the University of Wisconsin–Madison and a member of the Institute for Information Transmission Problems at the Russian Academy of Sciences. He is a leading researcher in the area of integrable probability, and has been awarded several prizes, including the Sloan Research Fellowship and the Prize of the Moscow Mathematical Society.

Lectures on Random Lozenge Tilings

VADIM GORIN

University of Wisconsin–Madison
Institute for Information Transmission Problems

CAMBRIDGE
UNIVERSITY PRESS

University Printing House, Cambridge CB2 8BS, United Kingdom

One Liberty Plaza, 20th Floor, New York, NY 10006, USA

477 Williamstown Road, Port Melbourne, VIC 3207, Australia

314–321, 3rd Floor, Plot 3, Splendor Forum, Jasola District Centre,
New Delhi – 110025, India

103 Penang Road, #05–06/07, Visioncrest Commercial, Singapore 238467

Cambridge University Press is part of the University of Cambridge.

It furthers the University's mission by disseminating knowledge in the pursuit of
education, learning, and research at the highest international levels of excellence.

www.cambridge.org
Information on this title: www.cambridge.org/9781108843966
DOI: 10.1017/9781108921183

First published 2021

A catalogue record for this publication is available from the British Library.

ISBN 978-1-108-84396-6 Hardback

Contents

Contents

Preface

These are lecture notes for a one-semester class devoted to the study of random tilings. It was 18.177 taught at the Massachusetts Institute of Technology during the spring of 2019. The brilliant students who participated in the class,[1] Andrew Ahn, Ganesh Ajjanagadde, Livingston Albritten, Morris (Jie Jun) Ang, Aaron Berger, Evan Chen, Cesar Cuenca, Yuzhou Gu, Kaarel Haenni, Sergei Korotkikh, Roger Van Peski, Mehtaab Sawhney, and Mihir Singhal, provided tremendous help in typing the notes.

Additional material was added to most of the lectures after the class was over. Hence, when using this review as a textbook for a class, one should not expect to cover all the material in one semester; something should be left out.

I would like to thank my wife, Anna Bykhovskaya, for her advice, love, and support. I also thank Amol Aggarwal, Alexei Borodin, Richard Kenyon, Christian Krattenthaler, Arno Kuijlaars, Igor Pak, Jiaming Xu, Marianna Russkikh, and Semen Shlosman for their useful comments and suggestions. I am grateful to Christophe Charlier, Maurice Duits, Sevak Mkrtchyan, and Leonid Petrov for the help with the simulations of random tilings.

Funding acknowledgments. The work of Vadim Gorin was partially supported by National Science Foundation (NSF) Grants DMS-1664619 and DMS-1949820, by the NEC Corporation Fund for Research in Computers and Communications, and by the Office of the Vice Chancellor for Research and Graduate Education at the University of Wisconsin–Madison with funding from the Wisconsin Alumni Research Foundation. Lectures 6 and 13 of this work were supported by the Russian Science Foundation (Project 20-41-09009).

---

[1] In alphabetical order by last name.

1

Lecture 1: Introduction and Tileability

1.1 Preamble

The goal of the lectures is for the reader to understand the mathematics of tilings. The general setup is to take a lattice domain and tile it with elementary blocks. For the most part, we study the special case of tiling a polygonal domain on the triangular grid (of mesh size 1) by three kinds of rhombi that we call "lozenges."

Panel (a) of Figure 1.1 shows an example of a polygonal domain on the triangular grid. Panel (b) of Figure 1.1 shows the lozenges: each of them is obtained by gluing two adjacent lattice triangles. A triangle of the grid is surrounded by three other triangles; attaching one of them, we get one of the three types of lozenges. The lozenges can also be viewed as orthogonal projections onto the $x + y + z = 0$ plane of three sides of a unit cube. Figure 1.2 provides an example of a lozenge tiling of the domain of Figure 1.1.

Figure 1.3 shows a lozenge tiling of a large domain, with the three types of lozenges shown in three different colors. The tiling here is generated uniformly at random over the space of all possible tilings of this domain. More precisely, it is generated by a computer that is assumed to have access to perfectly random bits. It is certainly not clear at this stage how such "perfect sampling" may be done computationally; in fact, we address this issue in the very last lecture. Figure 1.3 is meant to capture a "typical tiling," making sense of what this means is another topic that will be covered in this book. The simulation reveals an interesting feature: there are special regions next to the boundaries of the domain, and in each such region, there is only one (rather than three) type of lozenge. These regions are typically referred to as "frozen regions," and their boundaries are "arctic curves"; their discovery and study have been one of the important driving forces for investigations of the properties of random tilings.

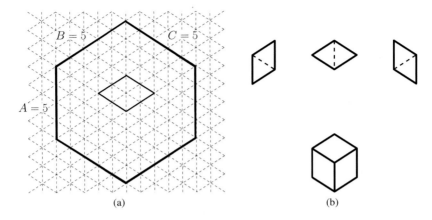

Figure 1.1 Panel (a): A $5 \times 5 \times 5$ hexagon with 2×2 rhombic hole. Panel (b): Three types of lozenges obtained by gluing two adjacent triangles of the grid.

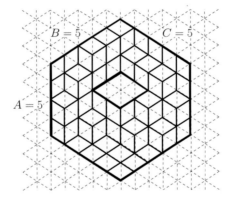

Figure 1.2 A lozenge tiling of a $5 \times 5 \times 5$ hexagon with a hole.

We often identify a tiling with a so-called height function. The idea is to think of a two-dimensional (2D) stepped surface living in a three-dimensional (3D) space and treat tiling as a projection of such surface onto $x + y + z = 0$ plane along the $(1, 1, 1)$ direction. In this way, three lozenges become projections of three elementary squares in 3D space parallel to each of the three coordinate planes. We formally define the height function later in this lecture. We refer to a web page of Borodin and Borodin[1] for a gallery of height functions in a 3D virtual-reality setting.

---

[1] A. Borodin and M. Borodin, A 3D representation for lozenge tilings of a hexagon, http://math.mit.edu/~borodin/hexagon.html.

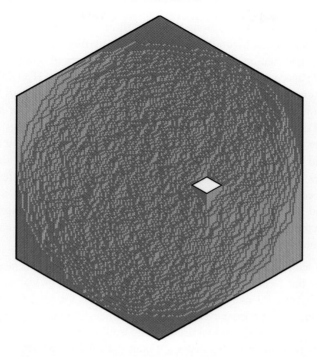

Figure 1.3 A perfect sample of a uniformly random tiling of a large hexagon with a hole. (I thank Leonid Petrov for this simulation.)

1.2 Motivation

Figure 1.3 is beautiful, and for mathematicians, this suffices for probing more deeply into it and trying to explain the various features that we observe in the simulation.

There are also some motivations from theoretical physics and statistical mechanics. Lozenge tilings serve as "toy models" that help in understanding the 3D Ising model (a standard model for magnetism). Configurations of the 3D Ising model are assignments of $\oplus$ and $\ominus$ spins to lattice points of a domain in $\mathbb{Z}^3$. A parameter (usually called "temperature") controls how much the adjacent spins are inclined to be oriented in the same direction. The zero-temperature limit leads to spins piling into as large as possible groups of the same orientation; these groups are separated with stepped surfaces whose projections in $(1, 1, 1)$ direction are lozenge tilings. For instance, if we start from the Ising model in a cube and fix boundary conditions to be $\oplus$ along three faces (sharing a single vertex) of this cube and $\ominus$ along the other three faces, then we end up with lozenge tilings of a hexagon in the zero-temperature limit.[2]

Another deformation of lozenge tilings is the six-vertex or square-ice model, whose state space consists of configurations of the molecules H_2O on the grid. There are six weights in this model (corresponding to the six ways to match an oxygen with two out of the neighboring four hydrogens), and for particular choices of the weights one discovers weighted bijections with tilings.

We refer the reader to Baxter (2007) for more information about the Ising model and the six-vertex model, further motivations to study them, and approaches to the analysis. In general, both the Ising and six-vertex models are more complicated objects than lozenge tilings, and they are much less understood. From this point of view, the theory of random tilings that we develop in these lectures can be treated as the first step toward the understanding of more complicated models of statistical mechanics.

For yet another motivation, we notice that the 2D stepped surfaces of our study have flat faces (these are frozen regions consisting of lozenges of one type, cf. Figures 1.3 and 1.5) and, thus, are relevant for modeling facets of crystals. One example from everyday life is a corner of a large box of salt. For a particular (nonuniform) random tiling model leading to the shapes reminiscent of such a corner, we refer the reader to Figure 10.1 in Lecture 10.

1.3 Mathematical Questions

We now turn to describing the basic questions that drive the mathematical study of tilings.

1. *Existence of tilings*: Given a domain $\mathcal{R}$ drawn on the triangular grid (and thus consisting of a finite family of triangles), does there exist a tiling of it? For example, a unit-sided hexagon is trivially tileable in two different ways, and the bottom part of Panel (b) in Figure 1.1 shows one of these tilings. On the other hand, if we take the equilateral triangle of side length 3 as our domain $\mathcal{R}$, then it is not tileable. This can be seen directly because the corner lozenges are fixed and immediately cause obstruction. Another way to prove nontileability is by coloring the unit triangles inside $\mathcal{R}$ in white and black colors in an alternating fashion. Each lozenge covers one black and one white triangle, but there is an unequal number of black and white triangles

---

[2] See Shlosman (2001), Cerf and Kenyon (2001), and Bodineau et al. (2005) for a discussion of the common features in low-temperature and zero-temperature 3D Ising models, as well as the interplay between the topics of this book and more classical statistical mechanics.

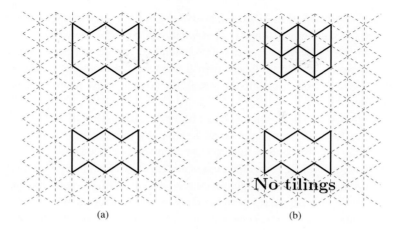

(a)   (b)

Figure 1.4 Panel (a): The top domain is tileable and the bottom one is not.
Panel (b): A possible tiling.

in the $\mathcal{R}$: the equilateral triangle of side length three has six triangles of one
color and three triangles of another color.

Another example is shown in Figure 1.4. In panel (a), we see two domains.
The top one is tileable, whereas the bottom one is not.

More generally, is there a simple (and efficient, from the computational
point of view) criterion for the tileability of $\mathcal{R}$? The answer is yes by a
theorem developed by Thurston (1990). We discuss this theorem in more
detail in Section 1.4 later in this lecture.

2. *How many tilings does a given domain $\mathcal{R}$ have?* The quality of the
answer depends on one's taste and perhaps what one means by a "closed-
form"/"explicit" answer. Here is one case where a "good" answer is known
by the following theorem due to MacMahon (who studied this problem in
the context of plane partitions, conjectured a formula in MacMahon (1896)
and proved it in the 1915 book, see Article 495 of MacMahon (1960)). Let
$\mathcal{R}$ be a hexagon with side lengths A, B, C, A, B, C in cyclic order. We denote
this henceforth by the $A \times B \times C$ hexagon (in particular, panel (a) of Figure
1.1 shows $5 \times 5 \times 5$ hexagon with a rhombic hole).

Theorem 1.1 (MacMahon, 1896)   *The number of lozenge tilings of $A \times B \times C$
hexagon equals*

$$\prod_{a=1}^{A}\prod_{b=1}^{B}\prod_{c=1}^{C}\frac{a+b+c-1}{a+b+c-2}. \tag{1.1}$$

As a sanity check, one can take $A = B = C = 1$, yielding the answer 2,
and indeed, one readily checks that there are precisely two tilings of $1 \times 1 \times 1$

hexagon. A proof of this theorem is given in Section 2.2. Another situation where the number of tilings is somewhat explicit is for the torus, and we discuss this in Lectures 3 and 4. For general $\mathcal{R}$, one cannot hope for such nice answers, yet certain determinantal formulas (involving large matrices whose size is proportional to the area of the domain) exist, as we discuss in Lecture 2.

3. *Law of large numbers*: Each lozenge tilings is a projection of a 2D surface and therefore can be represented as a graph of a function of two variables, which we call the "height function" (its construction is discussed in more detail in Section 1.4). If we take a *uniformly random* tiling of a given domain, then we obtain a random height function $h(x, y)$ encoding a random stepped surface. What is happening with the random height function of a domain of linear size L as $L \rightarrow \infty$? As we will see in Lectures 5–10 and in Lecture 23, the rescaled height function has a deterministic limit

$$\lim_{L \to \infty} \frac{1}{L} h(Lx, Ly) = \hat{h}(x, y).$$

An important question is how to compute and describe the *limit shape* $\hat{h}(x, y)$. One feature of the limit shapes of tilings is the presence of regions where the limiting height function is linear. In terms of random tilings, these are "frozen" regions, which contain only one type of lozenge. In particular, in Figure 1.3, there is a clear outer frozen region near each of the six vertices of the hexagon; another four frozen regions surround the hole in the middle.

Which regions are "liquid," that is, contain all three types of lozenges? What is the shape of the "arctic curve," that is, the boundary between frozen and liquid regions? For example, with the $aL \times bL \times cL$ hexagon setup, one can visually see from Figure 1.5 that the boundary appears to be an inscribed ellipse:

Theorem 1.2 (Baik et al., 2003; Cohn et al., 1998; Gorin, 2008; Petrov, 2014a)   *For $aL \times bL \times cL$ hexagon, a uniformly random tiling is with high probability asymptotically frozen outside the inscribed ellipse as $L \rightarrow \infty$. In more detail, for each (x, y) outside the ellipse, with probability tending to 1 as $L \rightarrow \infty$, all the lozenges that we observe in a finite neighborhood of (xL, yL) are of the same type.*

The inscribed ellipse of Theorem 1.2 is the unique degree 2 curve tangent to the hexagon's sides. This characterization in terms of algebraic curves extends to other polygonal domains, where one picks the degree such that there is a unique algebraic curve tangent (in the interior) of $\mathcal{R}$. Various approaches to Theorem 1.2, its relatives, and generalizations are discussed in Lectures 7, 10, 16, 21, 23.

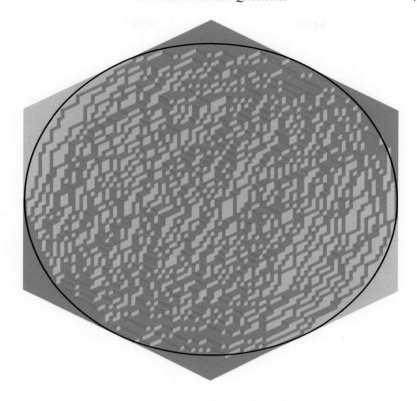

Figure 1.5 Arctic circle of a lozenge tiling.

4. *Analogs of the central limit theorem*: The next goal is to understand the random field of fluctuations of the height function around the asymptotic limit shape, that is, to identify the limit

$$\lim_{L \to \infty} (h(Lx, Ly) - \mathbb{E}[h(Lx, Ly)]) = \xi(x, y). \tag{1.2}$$

Note the unusual scaling; one may naively expect a need for dividing by $\sqrt{L}$ to account for fluctuations, as in the classical central limit theorem for sums of independent random variables and many similar statements. But there turns out to be some "rigidity" in tilings, and the fluctuations are much smaller. $\xi(x, y)$ denotes the limiting random field; in this case, it can be identified with the so-called "Gaussian free field." The Gaussian free field is related to conformal geometry because it turns out to be invariant under conformal transformations. This topic will be explored in Lectures 11,

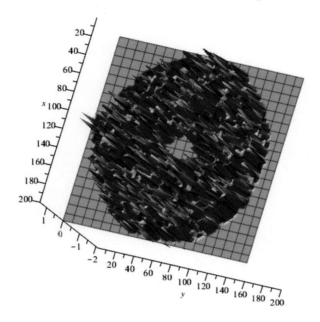

Figure 1.6 Fluctuations of the centered height function for lozenge tilings of a hexagon with a hole. Another drawing of the same system is shown in Figure 24.2 later in the text.

12, 21, and 23. For now, we confine ourselves to yet another picture, given in Figure 1.6.

5. *Height of a plateau/hole*: Consider Figures 1.2 and 1.3. The central hole has an integer-valued random height. What is the limiting distribution of this height? Note that comparing with (1.2), we expect that no normalization is necessary as $L \to \infty$, and therefore the distribution remains discrete as $L \to \infty$. Hence, the limit cannot be Gaussian. You can make a guess now or proceed to Lecture 24 for the detailed discussion.

6. *Local limit*: Suppose we "zoom in" at a particular location inside a random tiling of a huge domain. What are the characteristics of the tiling there? For example, consider a certain finite pattern of lozenges; call it $\mathcal{P}$, and see Figure 1.7 for an example. Asymptotically, what is the probability that $\mathcal{P}$ appears in the vicinity of (Lx, Ly)? Note that if $\mathcal{P}$ consists of a single lozenge, then we are just counting the local proportions for the lozenges of three types; hence, one can expect that they are reconstructed from the gradients of the limit shape $\tilde{h}$. However, for more general $\mathcal{P}$, it is not clear

Figure 1.7 An example of a local pattern $\mathcal{P}$ of lozenges. The bulk-limit question asks about the probability of observing such (or any other) pattern in a vicinity of a given point (Lx, Ly) in a random tiling of a domain of linear scale $L \to \infty$.

what to expect. This is called a "bulk-limit" problem, and we return to it in Lectures 16 and 17.

7. *Edge limit*: How does the arctic curve (border of the frozen region) fluctuate? What is the correct scaling? It turns out to be $L^{\frac{1}{3}}$ here, something that is certainly not obvious at all right now. The asymptotic law of rescaled fluctuations turns out to be given by the celebrated Tracy–Widom distribution from random matrix theory, as we discuss in Lectures 18 and 19.

8. *Sampling*: How does one sample from the uniform distribution over tilings? The number of tilings grows extremely fast (see, e.g., the MacMahon formula (1.1)), so one can not simply exhaustively enumerate the tilings on a computer, and a smarter procedure is needed. We discuss several approaches to sampling in Lecture 25.

9. *Open problem*: Can we extend the theory to 3D tiles?

1.4 Thurston's Theorem on Tileability

We begin our study from the first question: Given a domain $\mathcal{R}$, is there at least one tiling? The material here is essentially based on Thurston (1990).

Without loss of generality, we may assume $\mathcal{R}$ is a connected domain; the question of tileability of a domain is equivalent to that of its connected components. We start by assuming that $\mathcal{R}$ is simply connected, and then remove this restriction.

We first discuss the notion of a height function in more detail, and how it relates to the question of the tileability of a domain. There are six directions on the triangular grid, and the unit vectors in those directions are as follows:

$$a = (0, 1), \quad b = \left(-\frac{\sqrt{3}}{2}, -\frac{1}{2}\right), \quad c = \left(\frac{\sqrt{3}}{2}, -\frac{1}{2}\right), \quad -a, \quad -b, \quad -c.$$

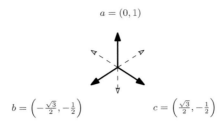

Figure 1.8 Out of the six lattice directions, three are chosen to be positive (in bold).

We call a, b, c positive directions, and their negations are negative directions, as in Figure 1.8.

We now define an *asymmetric* nonnegative distance function $d(u, v)$ for any two vertices u, v on the triangular grid (which is the lattice spanned by a and b) and a domain $\mathcal{R}$. $d(u, v)$ is the minimal number of edges in a positively oriented path from u to v staying within (or on the boundary of) $\mathcal{R}$. This is well defined as we assumed $\mathcal{R}$ to be connected. The asymmetry is clear: Consider $\mathcal{R}$ consisting of a single triangle, and let u, v be two vertices of it. Then $d(u, v) = 1$, and $d(v, u) = 2$ (or vice versa).

We now formally define a height function $h(v)$ for each vertex $v \in \mathcal{R}$, given a tiling of $\mathcal{R}$. This is given by a local rule: if $u \to v$ is a positive direction, then

$$h(v) - h(u) = \begin{cases} 1, & \text{if we follow an edge of a lozenge,} \\ -2, & \text{if we cross a lozenge diagonally.} \end{cases} \tag{1.3}$$

It may be easily checked that this height function is defined consistently. This is because the rules are consistent for a single lozenge (take, e.g., the lozenge $\{0, a, b, a + b\}$), and the definition extends consistently across unions. Note that h is determined up to a constant shift. We may assume without loss of generality that our favorite vertex v_0 has $h(v_0) = 0$.

Let us check that our definition matches the intuitive notion of the height. For that, we treat the positive directions a, b, c in Figure 1.8 as projections of coordinate axes Ox, Oy, Oz, respectively. Take one of the lozenges, say $\{0, a, b, a+b\}$. Up to rescaling by the factor $\sqrt{2/3}$, it can be treated as a projection of the square $\{(0, 0, 0), (1, 0, 0), (0, 1, 0), (1, 1, 0)\}$ onto the plane $x + y + z = 0$. Hence, our locally defined height function becomes the value of $x + y + z$. A similar observation is valid for two other types of lozenges. The conclusion is that if we identify a lozenge tiling with a stepped surface in 3D space, then our height is the (signed and rescaled) distance from the surface point to its projection onto the $x + y + z = $ const plane in the $(1, 1, 1)$ direction.

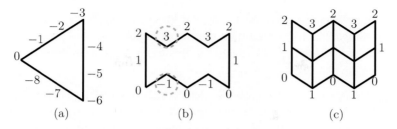

(a)   (b)   (c)

Figure 1.9 Three domains: the triangle in panel (a) has no tiles because there is no way to consistently define the heights along its boundary; the domain in panel (b) has no tilings because the inequality (1.4) fails for the encircled pair of points; the domain in panel (c) is tileable.

Note also that even if $\mathcal{R}$ is not tileable but is simply connected, then our local rules uniquely (up to global shift) define h on the boundary $\partial\mathcal{R}$: no lozenges are allowed to cross it, and therefore the increments in the positive direction all equal 1.

Exercise 1.3   Find the values of the height function along the two connected components of the boundary of the holey hexagon in panel (a) of Figure 1.1. Note that there are two arbitrary constants involved: one per connected component.

With these notions in hand, we state the theorem of Thurston (1990) on tileability.

Theorem 1.4   *Let $\mathcal{R}$ be a simply connected domain, with boundary $\partial\mathcal{R}$ on the triangular lattice. Then $\mathcal{R}$ is tileable if and only if both conditions hold:*

1. *One can define h on $\partial\mathcal{R}$, so that $h(v) - h(u) = 1$ whenever $u \to v$ is an edge of $\partial\mathcal{R}$, such that $v - u$ is a unit vector in one of the positive directions a, b, c.*
2. *This h satisfies*

$$\forall\, u, v \in \partial\mathcal{R}: \quad h(v) - h(u) \le d(u, v). \tag{1.4}$$

Before presenting the proof of Theorem 1.4, let us illustrate its conditions. For that, we consider the three domains shown in Figure 1.9, set $h = 0$ in the bottom-left corner of each domain, and further define h on $\partial\mathcal{R}$ by local rules following the boundary $\partial\mathcal{R}$ in the clockwise direction. For the triangle domain, there is no way to consistently define the height function along its boundary: when we circle around, the value does not match what we started from. And indeed, this domain is not tileable. For the middle domain of Figure 1.9, we can define the heights along the boundary, but (1.4) fails for the encircled points with height function values of -1 and 3. On the other hand, this domain clearly has no tilings. Finally, for the right domain of Figure 1.9, both conditions of Theorem 1.4 are satisfied; a (unique in this case) possible lozenge tiling is shown in the picture.

Proof of Theorem 1.4   Suppose $\mathcal{R}$ is tileable, and fix any tiling. Then it defines h on all of $\mathcal{R}$, including $\partial\mathcal{R}$, by local rules (1.3). Take any positively oriented path from u to v. Then the increments of d along this path are always 1, whereas the increments of h are either 1 or -2. Thus, (1.4) holds.

Now suppose that h satisfies (1.4). Define the following for $v \in \mathcal{R}$:

$$h(v) = \min_{u \in \partial\mathcal{R}} \left[d(u, v) + h(u) \right]. \tag{1.5}$$

We call this the *maximal height function* extending $h|_{\partial\mathcal{R}}$ and corresponding to the *maximal tiling*. Because d is nonnegative, the definition (1.5) of h matches on $\partial\mathcal{R}$ with the given in the theorem $h|_{\partial\mathcal{R}}$. On the other hand, because the inequality $h(v) - h(u) \le d(u, v)$ necessarily holds for any height function h and any $u, v \in \mathcal{R}$ (by the same argument as for the previous case of $u, v \in \partial\mathcal{R}$), no height function extending $h|_{\partial\mathcal{R}}$ can have a larger value at v than (1.5).

Claim. h defined by (1.5) changes by 1 or -2 along each positive edge.

The claim would allow us to reconstruct the tiling uniquely: each -2 edge gives a diagonal of a lozenge and hence a unique lozenge; each triangle has exactly one -2 edge because the only way to position increments $+1, -2$ along three positive edges is $+1 + 1 - 2 = 0$; hence, for each triangle, we uniquely reconstruct the lozenge to which it belongs. As such, we have reduced our task to establishing the claim.

Let $v \to w$ be a positively oriented edge in $\mathcal{R} \setminus \partial\mathcal{R}$. We begin by establishing some estimates. For any $u \in \partial\mathcal{R}$, $d(u, w) \le d(u, v) + 1$ by augmenting the u, v path by a single edge. Thus, we have

$$h(w) \le h(v) + 1.$$

Similarly, we may augment the u, w path by two positively oriented edges (at least one of the left/right pair across v, w is available) to establish $d(u, v) \le d(u, w) + 2$, and so

$$h(w) \ge h(v) - 2.$$

It remains to rule out the values 0 and -1 for $h(v) - h(w)$. For this, notice that any closed path on the triangular grid has a length divisible by 3. Thus, if the same boundary vertex $u \in \partial\mathcal{R}$ is involved in a minimizing path to v, w simultaneously, we see that $0, -1$ is ruled out by $h(w) \equiv h(v) + 1 \pmod 3$ because any other path to w has equivalent length modulo 3 to one of the "small augmentations" noted previously. Now suppose a different u' is involved in a minimizing path to w, whereas u is involved with a minimizing path to v. By looking at the closed loop $u, p_1(u, v), p_2(v, u'), p_\partial(u', u)$, where p_1 is a minimizing path from u to v, p_2 is the reversal of a minimizing path from w to u', augmented similar to the previous example, and p_∂ is a segment of the boundary $\partial\mathcal{R}$, we see that, once again, we are equivalent to one of the "small augmentations." Thus, in any case, $h(w) = h(v) + 1$ or $h(w) = h(v) - 2$, and we have completed the proof of the theorem. □

How can we generalize this when $\mathcal{R}$ is not simply connected? We note two key issues:

1. It is not clear how to define h on $\partial\mathcal{R}$ if one is not already given a tiling. On each closed-loop piece of $\partial\mathcal{R}$, we see that h is determined up to a constant shift, but these constants may differ across loops.
2. The step of the previous proof that shifts from u to u' used the fact that we could simply move along $\partial\mathcal{R}$. This is not, in general, true for a multiply connected domain.

These issues can be addressed by ensuring that $d(u, v) - h(v) + h(u) = 3k(u, v)$ for $k(u, v) \in \mathbb{N}$, $\forall u, v$, and simply leaving the ambiguity in h as it is:

Corollary 1.5 *Let $\mathcal{R}$ be a domain that is not simply connected, with boundary $\partial\mathcal{R}$ on the triangular lattice. Define h along $\partial\mathcal{R}$; this is uniquely defined up to constants $c_1, c_2, \ldots, c_l$, corresponding to constant shifts along the l pieces of $\partial\mathcal{R}$. Then $\mathcal{R}$ is tileable if and only if there exist $c_1, \ldots, c_l$ such that for every u and v in $\partial\mathcal{R}$:*

$$d(u, v) - h(v) + h(u) \geq 0, \tag{1.6a}$$
$$d(u, v) - h(v) + h(u) \equiv 0 \pmod 3. \tag{1.6b}$$

Proof of Corollary 1.5 ⇒: Equation (1.6a) ensures that this part of the proof of Theorem 1.4 remains valid; (1.6b) is also true when we have a tiling, again by the previous proof. ⇐: The shift from u to u' is now valid, so the proof carries through. □

We remark that Thurston's Theorem 1.4 and the associated height function method provide an $O(|\mathcal{R}| \ln(|\mathcal{R}|))$ time algorithm for tileability. There are recent improvements; for example, see Theorem 1.2 in Pak et al. (2016) for an $O(|\partial\mathcal{R}| \ln(|\partial\mathcal{R}|))$ algorithm in the simply connected case. Thiant (2003) provides an $O(|\mathcal{R}|l + a(\mathcal{R}))$ algorithm, where l denotes the number of holes in $\mathcal{R}$, and $a(\mathcal{R})$ denotes the area of all the holes of $\mathcal{R}$. We also refer the reader to

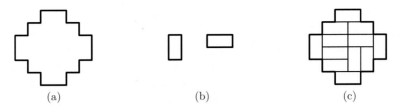

Figure 1.10 Panel (a): Aztec diamond domain on the square grid. Panel (b): Two types of dominoes. Panel (c): One possible domino tiling.

Section 5 of Pak et al. (2016) for further discussion of the optimal algorithms for tileability from the computer science literature.

1.5 Other Classes of Tilings and Reviews

Throughout these lectures, we are going to concentrate on lozenge tilings. However, the theory of random tilings is not restricted to only them.

The simplest possible alternative version of the theory deals with *domino tilings* on a square grid. In this setting, we consider a domain drawn on the square grid $\mathbb{Z}^2$ and its tilings with horizontal and vertical 2×1 rectangles called "dominoes." Figure 1.10 shows a domain known as the "Aztec diamond" and one of its domino tilings. Most of the results that we discuss in the book have their counterparts for domino tilings. For instance, their height functions and the question of tileability are discussed in the same article (Thurston, 1990) as for lozenges. The definitions become slightly more complicated, and we refer the reader to a web page of Borodin and Borodin[3] for an appealing 3D visualization of the height functions of domino tilings. Some aspects of the random tilings of the Aztec diamond are reviewed in Chapter 12 of Baik et al. (2016) and in Johansson (2016). Two other lecture notes on tilings that we are happy to recommend, Kenyon (2009) and Toninelli (2019), deal with dimers (or perfect matchings), which combine both lozenge and domino tilings.

Some additional classes of tilings and techniques for their enumeration can be found in Propp (2015). However, as we drift away from lozenges and dominos, the amount of available information about random tilings starts to decrease. For instance, *square triangle* tilings have central importance in representation theory and algebraic combinatorics because they appear in the enumeration of *Littlewood–Richardson coefficients*; see Purbhoo (2008) or Zinn-Justin (2009).

---

[3] A. Borodin and M. Borodin, A 3D representation for domino tilings of the Aztec diamond, http://math.mit.edu/~borodin/aztec.html.

Yet, at the time of writing this book, our understanding of the asymptotic behavior of random square triangle tilings is very limited.

Moving further away from the topics of this book, an introductory article (Ardila and Stanley, 2010) and two detailed textbooks (Golomb, 1995; Grünbaum and Shepard, 2016) show what else can be hiding under the word "tilings."

2

Lecture 2: Counting Tilings through Determinants

The goal of this lecture is to present two distinct approaches to the counting of lozenge tilings. Both yield different determinantal formulae for the number of tilings.

A statistical mechanics tradition defines a *partition function* as a total number of configurations in some model or, more generally, the sum of the weights of all configurations in some model. The usual notation is to use the capital letter Z in various fonts for the partition functions. In this language, we are interested in tools for the evaluation of the partition functions Z for tilings. In Lectures 2 and 3 we present several approaches to computing the partition functions, whereas Lecture 4 contains the first asymptotic results for the partition functions for tilings of large domains – we deal with large torus there. In addition to being interesting on their own, these computations will form a base for establishing the Law of Large Numbers and the Variational Principle for random tilings in subsequent Lectures 6, 7, and 8.

2.1 Approach 1: Kasteleyn Formula

The first approach relies on what may be called "Kasteleyn theory" (or more properly, "Kasteleyn–Temperley–Fisher (KTF) theory"), originally developed in Kasteleyn (1961) and Temperley and Fisher (1961). We illustrate the basic approach in Figure 2.1 for lozenge tilings on the triangular grid. One may color the triangles in two colors in an alternating fashion, similar to a checkerboard. Connecting the centers of adjacent triangles, one obtains a hexagonal *dual lattice* to the original triangular lattice, with black and white alternating vertices. The graph of this hexagonal dual is clearly bipartite with this coloring. Consider a simply connected domain $\mathcal{R}$ that consists of an equal number of triangles of two types (otherwise, it cannot be tiled); here, we simply use a unit hexagon. Then lozenge tilings of it are in bijection with perfect matchings of the associated bipartite graph ($\mathcal{G}$) formed by restriction of the dual lattice to

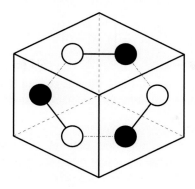

Figure 2.1 The hexagonal dual and the connection of tilings to perfect matchings.

the region corresponding to $\mathcal{R}$, here a hexagon. As such, our question can be rephrased as that of counting the number of perfect matchings of a (special) bipartite graph. The solid lines in Figure 2.1 illustrate one of the two perfect matchings (and hence tilings) in this case.

The basic approach of KTF is to relate the number of perfect matchings of the bipartite $\mathcal{G}$ to the determinant of a certain matrix K (*Kasteleyn matrix*) associated with $\mathcal{G}$. Let there be N black vertices and N white vertices. Then, K is an $N \times N$ matrix with entries as follows:

$$K_{ij} = \begin{cases} 1, & \text{if white } i \text{ is connected to black } j, \\ 0, & \text{otherwise.} \end{cases}$$

Then we have the following theorem:

Theorem 2.1 (Kasteleyn, 1961; Temperley and Fisher, 1961) *Let $\mathcal{R}$ be a simply connected domain on the triangular grid, and let $\mathcal{G}$ be its associated bipartite graph on the dual hexagonal grid. Then, the number of lozenge tilings of $\mathcal{R}$ is $|\det(K)|$.*

Before getting into the proof of this theorem, it is instructive to work out a simple example. Let $\mathcal{R}$ be a unit hexagon; label its vertices in the dual graph by $1, 1', 2, 2', 3, 3'$ in clockwise order. Then,

$$K = \begin{bmatrix} 1 & 0 & 1 \\ 1 & 1 & 0 \\ 0 & 1 & 1 \end{bmatrix}.$$

Then $\det(K) = 2$, which agrees with the number of tilings. However, also note that one can swap $1', 2$ and label the graph by $1, 2', 2, 1', 3, 3'$ in clockwise order, resulting in

$$K' = \begin{bmatrix} 0 & 1 & 1 \\ 1 & 1 & 0 \\ 1 & 0 & 1 \end{bmatrix},$$

and $\det(K') = -2$. Thus, the absolute value is needed. This boils down to the fact that there is no canonical order for the black vertices relative to the white ones.

Proof of Theorem 2.1 First, note that although there are $N!$ terms in the determinant, most of them vanish. More precisely, a term in the determinant is nonzero if and only if it corresponds to a perfect matching: using two edges adjacent to a single vertex corresponds to using two matrix elements from a single row/column and is not a part of the determinant's expansion.

The essence of the argument is thus understanding the signs of the terms. We claim that for simply connected $\mathcal{R}$ (the hypothesis right now; subsequent lectures will relax it), all the signs are the same. We remark that this is a crucial step of the theory; dealing with permanents (which are determinants without the signs) is far trickier if not infeasible.

Given its importance, we present two proofs of this claim:

1. The first one relies on the height function theory developed in Lecture 1. Define *elementary rotation E* that takes the matching of the unit hexagon $(1, 1'), (2, 2'), (3, 3')$ to $(1, 3'), (2, 1'), (3, 2')$ – that is, it swaps the solid and dash-dotted matchings in Figure 2.1. We may also define E^{-1}. Geometrically, E and E^{-1} correspond to removing and adding a single cube on the stepped surface (cf. Lecture 1). We claim that any two tilings of $\mathcal{R}$ are connected by a sequence of elementary moves of form E, E^{-1}. It suffices to show that we can move from any tiling to the *maximal tiling*, that is, the tiling corresponding to the point-wise maximal height function.[1] For that, geometrically, one can simply add one cube at a time until no more additions are possible. This process terminates on any simply connected domain $\mathcal{R}$, and it can only end at the maximal tiling.

   It now remains to check that E does not alter the sign of a perfect matching. It is clear that the sign of a perfect matching is just the number of inversions in the black-to-white permutation obtained from the matching (by definition of K and det). In the previous examples, these permutations

---

[1] The point-wise maximum of two height functions (that coincide at some point) is again a height function because the local rules (1.3) are preserved – for instance, one can show this by induction in the size of the domain. Hence, the set of height functions with fixed boundary conditions has a unique maximal element. In fact, we explicitly constructed this maximal element in our proof of Theorem 1.4.

are $\pi(1, 2, 3) = (1, 2, 3)$ and $\pi'(1, 2, 3) = (3, 1, 2)$, respectively. Note that $(3, 1, 2)$ is an even permutation, so composing with it does not alter the parity of the number of inversions (i.e., the parity of the permutation itself). Thus, all perfect matchings have the same sign.

2. The second approach relies on directly comparing two perfect matchings $\mathcal{M}_1$ and $\mathcal{M}_2$ of the same simply connected domain $\mathcal{R}$. We work on the dual hexagonal graph. Consider the union of these two matchings. Each vertex has degree 2 now (by the perfect matching hypothesis). Thus, the union consists of a bunch of doubled edges as well as loops. The doubled edges may be ignored (they correspond to common lozenges). $\mathcal{M}_1$ and $\mathcal{M}_2$ are obtained from each other by rotation along the loops, in a similar fashion to the operation E noted previously, except possibly across a larger number of edges. For a loop of length $2p$ (p black, p white), the sign of this operation is $(-1)^{p-1}$ because it corresponds to a cycle of length p that has $p - 1$ inversions. Thus, it suffices to prove that p is always odd.

   Here we use the specific nature of the hexagonal dual graph. First, note that each vertex has degree 3. Thus, any loop that does not repeat an edge cannot self-intersect at a vertex and is thus simple. Such a loop encloses some number of hexagons. We claim that p has opposite parity to the total (both black and white) number of vertices strictly inside the loop. We prove this claim by induction on the number of enclosed hexagons. With a single hexagon, $p = 3$, the number of vertices inside is 0, and the claim is trivial. Consider a contiguous domain made of hexagons $\mathcal{P}$ with a boundary of length $2p$. One can always remove one boundary hexagon such that it does not disconnect the domain $\mathcal{P}$. Doing a case analysis on the position of the surrounding hexagons, we see that the parity of the boundary loop length (measured in terms of, say, black vertices) remains opposite to that of the total number of interior vertices when we remove this hexagon. Hence, the claim follows by induction.

   It remains to note that the number of vertices inside each loop is even. Indeed, otherwise, there would have been no perfect matchings of the interior vertices.[2] Thus, p is always odd.   □

Remark 2.2   In general, the permanental formula for counting the perfect matchings of a graph is always valid. However, to get a determinantal formula, one needs to introduce signs/factors into K. It turns out that one can always find a consistent set of signs for counting the matchings of any *planar* bipartite graph. This is quite nontrivial and involves the *Kasteleyn orientation* of edges (Kasteleyn, 1963, 1967). The hexagonal case is simple because one can use

---

[2] The assumption of the domain being simply connected is used at this point.

the constant signs by the previous proof. For nonplanar graphs, good choices of signs are not known. However, for special cases, such as the torus that will be covered in Lecture 3, small modifications of the determinantal formula still work. More generally, on a genus g surface, the number of perfect matchings is given by a sum of 2^{2g} signed determinants; see Cimasoni and Reshetikhin (2007) and reference therein. For nonbipartite graphs, one needs to replace determinants with Pfaffians. Further information is available in, for instance, the work of Kasteleyn (1967), as well as the lecture notes on dimers by Kenyon (2009).

Exercise 2.3   Consider the domino tilings of the Aztec diamond, as in Figure 1.10. Find out what matrix elements we should take for the Kasteleyn matrix K so that its determinant gives the total number of tilings. (Hint: Try to use the 4th roots of unity: $1, -1, \mathbf{i}, -\mathbf{i}$).

Remark 2.4   The construction that we used in the second proof of Theorem 2.1 can be turned into an interesting stochastic system. Take two independent uniformly random lozenge tilings (equivalently, perfect matchings) of the same domain and superimpose them. This results in a collection of random loops that is known as the "double-dimer model." What is happening with this collection as the domain becomes large and the mesh size goes to 0? It is expected that one observes the *conformal loop ensemble* CLE_κ with $\kappa = 4$ in the limit. For tilings of general domains, this was not proven at the time when this book was written. However, partial results exist in the literature, and there is little doubt in the validity of this conjecture; see Kenyon (2014), Dubedat (2019), and Basok and Chelkak (2018).

2.2 Approach 2: Lindström–Gessel–Viennot Lemma

Suppose we want to apply the previously discussed machinery to derive MacMahon's formula for the number of tilings of an $\mathcal{R} = A \times B \times C$ hexagon. In principle, we have reduced the computation to that of a rather sparse determinant. However, it is not clear how we can proceed further. The goal of this section is to describe an alternative approach.

We use a bijection of tilings with another combinatorial object, namely, nonintersecting lattice paths, as illustrated in Figure 2.2.

We describe the bijection as follows: Enumerate (without loss, consistent with the previous orientation of the hexagon) the three fundamental lozenges l_a, l_b, l_c with their longer diagonals inclined at angles $0, -\pi/3, \pi/3$, respectively.

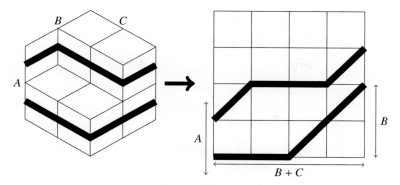

Figure 2.2 Bijection with nonintersecting lattice paths.

Leave l_a as is, and draw stripes at angles $-\pi/6, \pi/6$ connecting the midpoints of opposite sides of lozenges l_b, l_c, respectively. There is thus a bijection between lozenge tilings of the hexagon and nonintersecting lattice paths connecting the two vertical sides of length A. We may apply an affine transformation to obtain nonintersecting lattice paths on $\mathbb{Z}^2$ connecting $(0,0), \ldots, (0, A-1)$ with $(B+C, B), \ldots, (B+C, B+A-1)$ by elementary steps $\vec{v}_b \triangleq (1,0)$, $\vec{v}_c \triangleq (1,1)$, corresponding to the stripes across l_b, l_c, respectively. For general domains, one can still set up this bijection; the problem is that the starting and ending locations may no longer be contiguous, something that plays a key role in the derivation of MacMahon's formula by this approach.

Various forms of the following statement were used by many authors; see Karlin and McGregor (1959a, 1959b) and Gessel and Viennot (1985), with the most general result appearing in Lindström (1973).

Theorem 2.5 *The number of nonintersecting paths that start at $x_1 < x_2 < \cdots < x_N$ and end at $y_1 < y_2 < \cdots < y_N$ after T steps is as follows:*

$$\det\left[\binom{T}{y_i - x_j}\right]_{i,j=1}^{N}. \tag{2.1}$$

Proof of Theorem 2.5 First, notice that for any permutation σ of $[N] = \{1, 2, \ldots, N\}$, we see that

$$\prod_{j=1}^{N} \binom{T}{y_{\sigma(j)} - x(j)}$$

counts the total number of collections of paths linking x_j to $y_{\sigma(j)}$ $\forall j \in [N]$. The terms come with various signs, depending on the parity of σ. The goal is to show that the sum over the *entangled* paths is 0. By entangled, we mean any set of paths that are not nonintersecting. Let the collection of entangled paths be denoted $\mathcal{E}$. We construct an *involution* $f \colon \mathcal{E} \to \mathcal{E}$ such that $\mathbf{sign}(\sigma(f(P))) = -\mathbf{sign}(\sigma(P))$ for all $P \in \mathcal{E}$, where $\sigma(P)$ denotes the permutation corresponding to which x_i gets connected to which y_j, and $\mathbf{sign}(\sigma)$ denotes the parity of σ. This would complete the task, by the following elementary:

$$2 \sum_{P \in \mathcal{E}} \mathbf{sign}(\sigma(P)) = \sum_{P \in \mathcal{E}} \mathbf{sign}(\sigma(P)) + \sum_{P \in \mathcal{E}} \mathbf{sign}(\sigma(f(P)))$$
$$= \sum_{P \in \mathcal{E}} \mathbf{sign}(\sigma(P)) + \sum_{P \in \mathcal{E}} -\mathbf{sign}(\sigma(P))$$
$$= 0.$$

The involution is achieved by "tail-swapping." Care needs to be taken to ensure that it is well defined because there can be many intersections. We take the rightmost intersection points of two paths (if there is more than one intersection point with the same abscissa, then we take one with the largest ordinate) and swap their tails to the right from the intersection point. Thus, if before the involution we had these two paths linking x_i to y_j and $x_{i'}$ to $y_{i'}$, then after the involution x_i is linked to $y_{j'}$ and $x_{i'}$ is linked to y_j. If paths had no intersections, then we do nothing. This is an involution because the chosen intersection point remains the same when we iterate f, and swapping twice gets us back to where we started. Furthermore, the parity of σ changes when we apply f. We note that other choices of the involution are possible as long as they are well defined.

After cancellation, what we have left are nonintersecting paths. In the specific setting here, they can only arise from σ being the identity because any other σ will result in intersections. The identity is an even permutation, so we do not need to take absolute values here, unlike with the Kasteleyn formula. □

Exercise 2.6 For a collection of paths $\mathcal{E}$, let $w(\mathcal{E})$ denote the sum of the vertical coordinates of all vertices of all paths. Fix a parameter q, and using the same method, find a q-version of (2.1). You should get a $N \times N$ determinantal formula for the sum of $q^{w(\mathcal{E})}$ over all collections of nonintersecting paths starting at $x_1 < x_2 < \cdots < x_N$ and ending at $y_1 < y_2 < \cdots < y_N$ after T steps. At $q = 1$, the formula should match (2.1).

We remark that for a general domain, we still do not know how to compute the determinant (2.1). However, if either x_i or y_j consists of consecutive integers, we can evaluate this determinant in "closed form." This is true in the case of the $A \times B \times C$ hexagon, and so we now prove Theorem 1.1.

By Theorem 2.5 and the bijection with tilings we described, we have reduced our task to the computation of the following:

$$\det_{1 \le i,j \le A} \begin{pmatrix} B+C \\ B-i+j \end{pmatrix}. \tag{2.2}$$

Our proof relies on the following lemma, which can be found in a very helpful reference for the evaluation of determinants (Krattenthaler, 1999a); for its earlier proof see also Lemma 2.2 in (Krattenthaler, 1990):

Lemma 2.7 *Let $X_1, \ldots, X_n, A_2, \ldots, A_n, B_2, \ldots, B_n$ be indeterminates. Then,*

$$\det_{1 \le i,j \le n} ((X_i + A_n)(X_i + A_{n-1}) \cdots (X_i + A_{j+1})(X_i + B_j)(X_i + B_{j-1}) \cdots (X_i + B_2))$$

$$= \prod_{1 \le i < j \le n} (X_i - X_j) \prod_{2 \le i \le j \le n} (B_i - A_j).$$

Proof   The proof is based on reduction to the standard Vandermonde determinant by column operations. First, subtract the $(n-1)$th column from the nth, the $(n-2)$th from the $(n-1)$th, $\ldots$, the first column from the second, to reduce the left-hand side to the following:

$$\left[\prod_{i=2}^{n} (B_i - A_i) \right] \det_{1 \le i,j \le n} ((X_i + A_n)(X_i + A_{n-1}) \cdots (X_i + A_{j+1})(X_i + B_{j-1}) \cdots (X_i + B_2)). \tag{2.3}$$

Next, repeat the same process with the determinant of (2.3), factoring out

$$\prod_{i=2}^{n-1} (B_i - A_{i+1}).$$

We can clearly keep repeating the process until we have reached the simplified form:

$$\left[\prod_{2 \le i \le j \le n} (B_i - A_j) \right] \det_{1 \le i,j \le n} ((X_i + A_n)(X_i + A_{n-1}) \cdots (X_i + A_{j+1})).$$

At this stage, we have a slightly generalized Vandermonde determinant, which evaluates to the following, as desired:[3]

$$\prod_{1 \le i < j \le n} (X_i - X_j). \qquad \square$$

---

[3] Here is a simple way to prove the last determinant evaluation: The determinant is a polynomial in X_i of degree $n(n-1)/2$. It vanishes whenever $X_i = X_j$, and hence it is divisible by each factor $(X_i - X_j)$. We conclude that the determinant is $C \times \prod_{i<j}(X_i - X_j)$, and it remains to compare the leading coefficients to conclude that $C = 1$.

Proof of Theorem 1.1   Observe that

$$\det_{1\le i,j\le A}\binom{B+C}{B-i+j} = \left[\prod_{i=1}^{A}\frac{(B+C)!}{(B-i+A)!(C+i-1)!}\right]$$

$$\times \det_{1\le i,j\le A}\left(\frac{(B-i+A)!\,(C+i-1)!}{(B-i+j)!\,(C+i-j)!}\right) = (-1)^{\binom{A}{2}}\left[\prod_{i=1}^{A}\frac{(B+C)!}{(B-i+A)!(C+i-1)!}\right]$$

$$\times \det_{1\le i,j\le A}\left[(i-B-A)(i-B-A+1)\cdots(i-B-j-1)\right.$$

$$\left.\times\,(i+C-j+1)(i+C-j+2)\cdots(i+C-1)\right].$$

Now take $X_i = i$, $A_j = -B-j$, $B_j = C-j+1$ in Lemma 2.7 to simplify further. We get the following:

$$(-1)^{\binom{A}{2}}\left[\prod_{i=1}^{A}\frac{(B+C)!}{(B-i+A)!(C+i-1)!}\right]\left[\prod_{1\le i<j\le A}(i-j)\right]\left[\prod_{2\le i\le j\le A}(C+B+1-i+j)\right]$$

$$= \left[\prod_{i=1}^{A}\frac{(B+C)!}{(B-i+A)!(C+i-1)!}\right]\left[\prod_{1\le i<j\le A}(j-i)\right]\left[\prod_{2\le i\le j\le A}(C+B+1-i+j)\right]$$

$$= \left[\prod_{i=1}^{A}\frac{(B+C)!}{(B-i+A)!(C+i-1)!}\right]\left[\prod_{1\le j<A}j!\right]\left[\prod_{2\le j\le A}\frac{(C+B+j-1)!}{(B+C)!}\right]$$

$$= \left[\prod_{i=2}^{A}\frac{1}{(B-i+A)!(C+i-1)!}\right]\left[\prod_{1\le j<A}j!\right]\left[\prod_{2\le j\le A}(C+B+j-1)!\right]\frac{(B+C)!}{(B+A-1)!C!}.$$

$$(2.4)$$

Perhaps the easiest way to get MacMahon's formula out of this is to induct on A. This is somewhat unsatisfactory because it requires knowledge of MacMahon's formula a priori, though we remark that this approach is common.

First, consider the base case $A = 1$. Then MacMahon's formula is

$$\prod_{b=1}^{B}\prod_{c=1}^{C}\frac{b+c}{b+c-1} = \prod_{b=1}^{B}\frac{b+C}{b}$$

$$= \binom{B+C}{B}.$$

This is the same as the expression (2.4) because all the explicit products are empty. Keeping B, C fixed but changing $A \to A + 1$, MacMahon's formula multiplies by the following factor:

$$\prod_{b=1}^{B}\prod_{c=1}^{C}\frac{A+b+c}{A+b+c-1} = \prod_{b=1}^{B}\frac{A+b+C}{A+b}$$

$$= \frac{(A+B+C)!}{(A+C)!}\frac{A!}{(A+B)!}.$$

Let us now look at the factor for (2.4). The numerator factors $(A+B+C)!$ and $A!$ arise from the right and middle explicit products in (2.4). The denominator factor $(A+C)!$ arises from the denominator term $(C+i-1)!$ of (2.4). The only remaining unaccounted part of the denominator of (2.4) that varies with A is $\prod_{i=1}^{A}(1/(B+A-i)!)$, which thus inserts the requisite $(A+B)!$ in the denominator on $A \to A+1$. This finishes the induction and hence the proof. $\square$

2.3 Other Exact Enumeration Results

There is a large collection of beautiful results in the literature giving compact closed formulas for the numbers of lozenge tilings[4] of various specific domains. There is no unifying guiding principle to identify the domains for which such formulas are possible, and there is always lots of intuition and guesswork involved in finding new domains (as well as numerous computer experiments with finding the numbers and attempting to factorize them into small factors).

Once a formula for the number of tilings of some domain is guessed, a popular way for checking it is to proceed by induction, using the *Dodgson condensation* approach for recursive computations of determinants. This approach relies on the Desnanot–Jacobi identity, a quadratic relation between the determinant of an $N \times N$ matrix and its minors of sizes $(N-1) \times (N-1)$ and $(N-2) \times (N-2)$. The combinatorial version of this approach is known as "Kuo condensation" (see, e.g., Ciucu, 2015 and references therein). For instance, Kratten-thaler (1999a) (following Zeilberger, 1996) proves MacMahon's formula of Theorem 1.1 in this way. Numerous generalizations of MacMahon's formula, including, in particular, exact counts for tilings with different kinds of symmetries, are reviewed in Krattenthaler (2016). One tool that turns out to be very useful in the enumeration of symmetric tilings is the matching factorization theorem of Ciucu (1997).

---

[4] Although we only discuss lozenge tilings in this section, there are many other fascinating exact enumerations in related models. Examples include simple formulas for the number of domino tilings of a rectangle in Temperley and Fisher (1961) and Kasteleyn (1961) and of the Aztec diamond in Elkies et al. (1992), as well as a determinantal formula for the partition function of the six-vertex model with domain-wall boundary conditions in Izergin (1987) and Korepin (1982).

Among other results, there is a large scope of literature devoted to the exact enumeration of lozenge tilings of hexagons with various defects; for example, see Krattenthaler (2002), Ciucu and Krattenthaler (2002), Ciucu and Fischer (2015), Lai (2017), Ciucu (2018), Rosengren (2016), and many more references cited therein. Ciucu (2008) further used formulas of this kind to emphasize the asymptotic dependence of tiling counts on the positions of defects, which resembles the laws of electrostatics.

Some of the enumeration results can be extended to the explicit evaluations of weighted sums over lozenge tilings, and we refer the reader to Borodin et al. (2010), Young (2010), and Morales et al. (2019) for several examples.

3

Lecture 3: Extensions of the Kasteleyn Theorem

3.1 Weighted Counting

In the previous lecture, we considered tilings of simply connected regions drawn on the triangular grid by lozenges. We saw that if we checkerboard-color the resulting triangles black and white, this corresponds to perfect matchings of a bipartite graph $G = (W \sqcup B, E)$. In this section, we extend the enumeration of matchings to the *weighted* situation.

The notational setup is as follows: Number the white and black vertices of G by $1, 2, \ldots, n$. We regard the region $\mathcal{R}$ as being equipped with a *weight* function $\mathbf{w}(\bullet) > 0$, assigning to each edge $ij \in E$ of G with white i and black j some positive weight. Then we let the Kasteleyn matrix $K_{\mathcal{R}}$ be an $n \times n$ matrix with entries

$$K_{\mathcal{R}}(i, j) = \begin{cases} \mathbf{w}(ij), & ij \text{ is an edge of } G, \\ 0, & \text{otherwise.} \end{cases}$$

The difference from the setting of the previous lecture is that $\mathbf{w}(\cdot)$ was identical to 1 there. When there is no risk of confusion, we abbreviate $K_{\mathcal{R}}$ to K. This matrix depends on the labeling of the vertices, but only up to the permutation of the rows and columns, and hence we will not be concerned with this distinction.

By abuse of notation, we will then denote the *weight* of a tiling T by

$$\mathbf{w}(T) \stackrel{\text{def}}{=} \prod_{\ell \in T} \mathbf{w}(\ell).$$

An extension of Theorem 2.1 states that we can compute the weighted sum of tilings as the determinant of the Kasteleyn matrix.

Theorem 3.1 *The weighted number of tilings of a simply connected domain $\mathcal{R}$ is equal to $\sum_T \mathbf{w}(T) = |\det K_{\mathcal{R}}|$.*

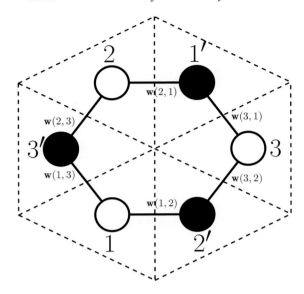

Figure 3.1 A unit hexagon with weights of edges for the corresponding graph.

The proof of Theorem 3.1 is the same as the unweighted case of counting the number of tilings (obtained by setting $\mathbf{w}(ij) = 1$ for each edge $ij \in E$) in Theorem 2.1. Let us remind the reader that the simply connected nature of $\mathcal{R}$ is used in the proof in order to show that the signs of all the terms in the determinant expansion are the same.

Example 3.2 (Kasteleyn Matrix of a Hexagon)   *The unit hexagon with six triangles can be modeled by a cycle graph G on six vertices, as shown in Figure 3.1.*
    Then

$$\det K_{\mathcal{R}} = \det \begin{bmatrix} 0 & \mathbf{w}(1,2) & \mathbf{w}(1,3) \\ \mathbf{w}(2,1) & 0 & \mathbf{w}(2,3) \\ \mathbf{w}(3,1) & \mathbf{w}(3,2) & 0 \end{bmatrix}$$

$$= \mathbf{w}(1,2)\mathbf{w}(2,3)\mathbf{w}(3,1) + \mathbf{w}(2,1)\mathbf{w}(3,2)\mathbf{w}(1,3),$$

and as advertised, the nonzero terms have the same sign and correspond to the weights of the two obvious tilings. Note that $\det K_{\mathcal{R}}$ is only defined up to sign because if we relabel the vertices in a different way (equivalently, change some rows and columns), then the determinant's sign changes.

3.2 Tileable Holes and Correlation Functions

There are situations in which the result of Theorem 3.1 is still true, even if $\mathcal{R}$ is not simply connected.

Proposition 3.3 *Suppose a region $\mathcal{R}$ is a difference of a simply connected domain and a union of disjoint lozenges inside this domain. Let K be the Kasteleyn matrix for $\mathcal{R}$. Then the weighted sum of perfect matchings of $\mathcal{R}$ equals $|\det K|$.*

Proof Let $\widetilde{\mathcal{R}}$ denote the simply connected region obtained by adding in the removed lozenges. Then every nonzero term in $\det K_{\mathcal{R}}$ can be mapped into a corresponding nonzero term in $\det K_{\widetilde{\mathcal{R}}}$ by adding in the deleted lozenges. We then quote the result that all terms in $\det K_{\widetilde{\mathcal{R}}}$ have the same sign, so the corresponding terms in $\det K_{\mathcal{R}}$ should all have the same sign, too. $\square$

There is an important probabilistic corollary of Proposition 3.3. Let $\mathcal{R}$ be as in the proposition, and fix a weight function $\mathbf{w}(\bullet) > 0$. We can then speak about *random tilings* by setting the probability of a tiling T to be

$$\mathbb{P}\,(\text{tiling } T) = \frac{1}{\mathcal{Z}}\mathbf{w}(T) = \frac{1}{\mathcal{Z}} \prod_{\text{lozenge } \ell \,\in T} \mathbf{w}(\ell).$$

The normalizing constant $\mathcal{Z} = \sum_T \mathbf{w}(T)$ is called *the partition function*.

Definition 3.4 Define the *nth correlation function* ρ_n as follows: given lozenges $\ell_1, \ldots, \ell_n$ we set

$$\rho_n(\ell_1, \ldots, \ell_n) = \mathbb{P}\,(\ell_1 \in T,\, \ell_2 \in T,\, \ldots, \ell_n \in T)\,.$$

The following proposition gives a formula for ρ_n:

Theorem 3.5 *Write each lozenge as $\ell_i = (w_i, b_i)$ for $i = 1, \ldots, n$. Then*

$$\rho_n(\ell_1, \ldots, \ell_n) = \prod_{i=1}^{n} \mathbf{w}(w_i, b_i) \det_{i,j=1,\ldots,n} \left[K^{-1}(b_i, w_j) \right]. \qquad (3.1)$$

Remark 3.6 The proposition in this form was stated in Kenyon (1997). However, the importance of the inverse Kasteleyn matrix has been known since the 1960s; see Montroll et al. (1963) and McCoy and Wu (1973).

Proof of Theorem 3.5   Using Proposition 3.3, we have

$$
\rho_n (\ell_1, \ldots, \ell_n) = \frac{1}{\mathcal{Z}} \sum_{\substack{\text{tiling } T \\ T \ni \ell_1, \ldots, \ell_n}} \prod_{\ell \in T} \mathbf{w}(\ell)
$$

$$
= \frac{1}{\mathcal{Z}} \prod_{i=1}^{n} \mathbf{w}(w_i, b_i) \left| \det_{i', j' \in \mathcal{R} \backslash \{\ell_1, \ldots, \ell_n\}} \left[K(w_{i'}, b_{j'}) \right] \right|
$$

$$
= \prod_{i=1}^{n} \mathbf{w}(w_i, b_i) \left| \frac{\displaystyle\det_{i', j' \in \mathcal{R} \backslash \{\ell_1, \ldots, \ell_n\}} \left[K(w_{i'}, b_{j'}) \right]}{\displaystyle\det_{i', j' \in \mathcal{R}} \left[K(w_{i'}, b_{j'}) \right]} \right|
$$

$$
= \prod_{i=1}^{n} \mathbf{w}(w_i, b_i) \det_{i, j = 1, \ldots, n} \left[K^{-1}(b_i, w_j) \right],
$$

where the last equality uses the generalized Cramer's rule (see, e.g., Prasolov, 1994, Section 2.5.2), which claims that a minor of a matrix is equal (up to sign) to the product of the complementary–transpose minor of the inverse matrix and the determinant of the original matrix. In particular, the $n = 1$ case of this statement is the computation of the inverse matrix as the transpose cofactor matrix divided by the determinant

$$
K^{-1}(b_i, w_j) = (-1)^{i+j} \frac{\displaystyle\det_{i', j' \in \mathcal{R} \backslash \{b_i, w_j\}} [K(w_{i'}, b_{j'})]}{\displaystyle\det_{i', j' \in \mathcal{R}} [K(w_{i'}, b_{j'})]}.
$$

Note that Proposition 3.3 involves the absolute value of the determinant. We leave it to the reader to check that the signs in the computation match and (3.1) has no absolute value.   □

3.3 Tilings on a Torus

Our next stop is to count tilings on the torus. The main motivation comes from the fact that the translation invariance of the torus allows us to use Fourier analysis to compute the determinants that Kasteleyn theory outputs – this will be important for the subsequent asymptotic analysis. Yet, for nonplanar domains (e.g., a torus), the Kasteleyn theorem needs a modification.

3.3.1 Setup

We consider a hexagonal grid on a torus $\mathbb{T} = S^1 \times S^1$, again with the corresponding bipartite graph $G = (W \sqcup B, E)$. The torus has a fundamental domain $\mathcal{F}$ drawn as a rhombus with n_1 and n_2 side lengths, as shown in Figure 3.2. We impose coordinates so that the black points are of the form

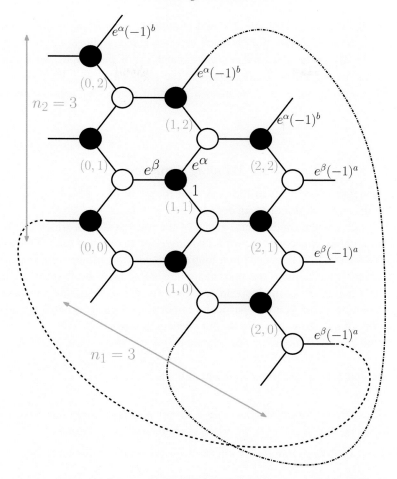

Figure 3.2 An $n_1 \times n_2 = 3 \times 3$ torus and its coordinate system. Three types of
edges have weights e^α, e^β, and 1. When we loop around the torus, the weights
get additional factors $(-1)^a$ or $(-1)^b$.

(x, y), where $0 \le x < n_1$ and $0 \le y < n_2$; see Figure 3.2. The white points have
similar coordinates, so the black and white points with the same coordinate are
linked by a diagonal edge, with the white vertex below. Our goal is to compute
the weighted number of tilings with a general weight function $\mathbf{w}(\bullet)$.

As with the original Kasteleyn theorem, perm K gives the number of perfect
matchings, and our aim is to compute the signs of the terms when the permanent
is replaced by the determinant.

To this end, we fix M_0, the matching using only diagonal edges between
vertices of the same coordinates – its edges are adjacent to coordinate labels

$((0, 0), (0, 1),$ etc.) in Figure 3.2 – and choose the numbering of the white and black vertices so that the ith black vertex is directly above the ith white vertex. We let M be a second arbitrary matching of the graph. We would like to understand the difference between the signs of M and M_0 in det K. As before, overlaying M and M_0 gives a 2-regular graph (i.e., a collection of cycles).

3.3.2 Winding Numbers

The topological properties of the cycles (or loops) in $M \cup M_0$ play a role in finding the signs of M and M_0 in det M.

We recall that the fundamental group of the torus is $\mathbb{Z} \times \mathbb{Z}$, which can be seen through the concept of the *winding number* of an oriented loop, which is a pair $(u, v) \in \mathbb{Z} \times \mathbb{Z}$ associated to a loop that counts the number of times the loop traverses the torus in two coordinate directions. In order to speak about the winding numbers of the loops in $M \cup M_0$, we need to orient them in some way. This orientation is not of particular importance because eventually, only the oddity of u and v will matter for our sign computations.

For the nontrivial loops, we need the following two topological facts:

Proposition 3.7   *[Facts from Topology] On the torus,*

- *a loop that is not self-intersecting and not null-homotopic has winding number $(u, v) \in \mathbb{Z} \times \mathbb{Z}$ with the greatest common divisor $\gcd(u, v) = 1$;*
- *any two such loops that do not intersect have the same winding number.*

Both facts can be proven by lifting the loops to $\mathbb{R}^2$ – the universal cover of the torus – and analyzing the resulting curves there, and we will not provide more details here.

Proposition 3.8   *Take all loops from $M \cup M_0$ that are not double edges. Then each such loop intersects the vertical border of the fundamental domain (of Figure 3.2) u times and the horizontal (diagonal) border of the fundamental domain v times for some (u, v) independent of the choice of the loop and with $\gcd(u, v) = 1$. Further, the length of the loop (i.e., the number of black vertices on it) is $n_1 u + n_2 v$.*

Proof Let γ denote one of the loops in question, and let us lift it to a path in $\mathbb{R}^2$ via the universal cover

$$\mathbb{R}^2 \twoheadrightarrow S^1 \times S^1.$$

We thus get a path γ' in $\mathbb{R}^2$ linking a point (x, y) to $(x + n_1 u', y + n_2 v')$ for some integers u' and v'. By Proposition 3.7, $\gcd(u', v') = 1$ (unless $u' = v' = 0$) and u', v' are the same for all loops. Let us orient γ' by requiring that all the edges coming from M_0 are oriented from black to white vertex. Then, because every second edge of γ' comes from M_0, we conclude that the x-coordinate is increasing and the y-coordinate is decreasing along the path γ'; that is, the steps of this path of one oddity are $(x', y') \to (x', y')$, and the steps of this path of another oddity are

$$(x', y') \to (x + 1, y) \qquad \text{or} \qquad (x', y') \to (x, y - 1), \tag{3.2}$$

in the coordinate system of Figure 3.2. The monotonicity of path γ' implies that $u' = v' = 0$ is impossible; that is, a nontrivial loop obtained in this way cannot be null-homotopic. The same monotonicity implies $u = |u'|$, $v = |v'|$ and that the length of the path is $|n_1 u'| + |n_2 v'|$. □

3.3.3 Kasteleyn Theorem on the Torus

In order to handle the wrap around caused by the winding numbers u and v, we have to modify our Kasteleyn matrix slightly.

Fix the fundamental domain $\mathcal{F}$ as in Figure 3.2. We have the following definition:

Definition 3.9 Let $(a, b) \in \{0, 1\}^2$. The Kasteleyn matrix $K_{a,b}$ is the same as the original K, except for the following:

- For any edge crossing the vertical side of $\mathcal{F}$ with n_1 vertices, we multiply its entry by $(-1)^a$.
- For any edge crossing the horizontal/diagonal side of $\mathcal{F}$ with n_2 vertices, we multiply its entry by $(-1)^b$.

Theorem 3.10 *For perfect matchings on the $n_1 \times n_2$ torus, we have*

$$\sum_T \mathbf{w}(T) = \frac{\varepsilon_{00} \det K_{00} + \varepsilon_{01} \det K_{01} + \varepsilon_{10} \det K_{10} + \varepsilon_{11} \det K_{11}}{2} \tag{3.3}$$

for some $\varepsilon_{ab} \in \{-1, 1\}$ which depends only on the parity of n_1 and n_2.

Proof   The correct choice of ε_{ab} is given by the following table:

$(n_1 \bmod 2, n_2 \bmod 2)$	$(0,0)$	$(0,1)$	$(1,0)$	$(1,1)$
ε_{00}	-1	$+1$	$+1$	$+1$
ε_{01}	$+1$	-1	$+1$	$+1$
ε_{10}	$+1$	$+1$	-1	$+1$
ε_{11}	$+1$	$+1$	$+1$	-1

(3.4)

For the proof, we note that each $\det K_{ab}$ is expanded as a signed sum of weights of tilings, and we would like to control the signs in this sum. For the matching M_0, the sign in all four K_{ab} is $+1$. Because the sum over each column in (3.4) is 2, this implies that the weight of M_0 enters into the right-hand side of (3.3) with the desired coefficient 1.

Any other matching M differs from M_0 by rotations along k loops of class (u, v), as described in Proposition 3.8. Note that at least one of the numbers u, v, should be odd because $\gcd(u, v) = 1$. The loop has the length $n_1 u + n_2 v$, and rotation along this loop contributes to the expansion of $\det K_{ab}$ the sign $(-1)^{n_1 u + n_2 v + 1 + au + bv}$. Thus, we want to show that the choice of ε_{ab} obeys

$$\sum_{a=0}^{1} \sum_{b=0}^{1} \varepsilon_{ab} \cdot (-1)^{k(n_1 u + n_2 v + au + bv + 1)} = 2.$$

If k is even, then all the signs are $+1$, and the check is the same as for M_0. For the odd k case, we can assume without loss of generality that $k = 1$, and we need to show that

$$\sum_{a=0}^{1} \sum_{b=0}^{1} \varepsilon_{ab} \cdot (-1)^{n_1 u + n_2 v + au + bv} = -2.$$

We use the fact that u, v are not both even to verify that the choice of (3.4) works. For example, if n_1 and n_2 are both even, then we get the sum

$$-1 + (-1)^u + (-1)^v + (-1)^{u+v}.$$

Checking the choices $(u, v) = (0, 1), (1, 0), (1, 1)$, we see that the sum is -2 in all the cases. For other oddities of n_1 and n_2, the proof is the same.   □

Remark 3.11   For the perfect matchings on a genus g surface, one would need to consider a signed sum of 2^{2g} determinants for counting, as first noted by Kasteleyn (1967). For detailed information on the correct choice of signs, see Cimasoni and Reshetikhin (2007) and references therein.

Exercise 3.12   Find an analogue of Theorem 3.10 for perfect matchings on an $n_1 \times n_2$ cylinder.

4

Lecture 4: Counting Tilings on a Large Torus

This lecture is devoted to the asymptotic analysis of the result of Theorem 3.10. Throughout this section, we use the e^α, e^β weight for tilings, as in Figure 3.2, with $a = b = 0$.

4.1 Free Energy

From the previous lecture, we know that on a torus with side lengths n_1 and n_2, the partition function is

$$Z(n_1, n_2) = \sum_{\text{Tilings}} \prod \mathbf{w}(\text{lozenges})$$

$$= \frac{\pm K_{0,0} + \pm K_{1,0} + \pm K_{0,1} + \pm K_{1,1}}{2},$$

where $K_{a,b}$ are the appropriate Kasteleyn matrices. Note that we can view $K_{a,b}$ as linear maps from $\mathbb{C}^B$ to $\mathbb{C}^W$, where B and W are the set of black and white vertices on the torus.

Our first task is to compute the determinants $\det K_{ab}$. The answer is explicit in the case of a translation-invariant weight function $\mathbf{w}$. Choose any real numbers α and β, and write

$$\mathbf{w}(ij) \in \left\{ 1, e^\alpha, e^\beta \right\},$$

according to the orientation of edge ij as in Figure 3.2. We evaluate

$$\det K_{ab} = \prod \text{eigenvalues } K_{ab}.$$

We start with K_{00} and note that it commutes with shifts in both directions. Because the eigenvectors of shifts are exponents, we conclude that so should be the eigenvectors of K_{00} (commuting family of operators can be diagonalized simultaneously). The eigenvalues are then computed explicitly, as summarized in the following claim:

Claim 4.1   *The eigenvectors of K_{00} are given as follows: given input $(x, y) \in \{0, \ldots, n_1 - 1\} \times \{0, \ldots, n_2 - 1\}$, the eigenvector is*

$$(x, y) \mapsto \exp\left(\mathbf{i} \cdot \frac{2\pi k}{n_1} x + \mathbf{i} \cdot \frac{2\pi \ell}{n_2} y\right)$$

for each $0 \le k < n_1$, $0 \le \ell < n_2$. The corresponding eigenvalue is

$$1 + e^{\beta} \exp\left(-\mathbf{i}\frac{2\pi k}{n_1}\right) + e^{\alpha} \exp\left(\mathbf{i}\frac{2\pi \ell}{n_2}\right).$$

Therefore, the determinant of $K_{0,0}$ is

$$\prod_{k=1}^{n_1}\prod_{\ell=1}^{n_2}\left(1 + e^{\beta} \exp\left(-\mathbf{i}\frac{2\pi k}{n_1}\right) + e^{\alpha} \exp\left(\mathbf{i}\frac{2\pi \ell}{n_2}\right)\right).$$

For $K_{a,b}$, we can prove the analogous facts essentially by perturbing the eigenvalues we obtained in the $K_{0,0}$ case. The proof is straightforward, and we omit it here.

Claim 4.2   *The eigenvectors of $K_{a,b}$ are*

$$(x, y) \mapsto \exp\left(\mathbf{i} \cdot \pi \frac{2k + a}{n_1} x + \mathbf{i} \cdot \pi \frac{2\ell + b}{n_2} y\right),$$

and the determinant of $K_{a,b}$ is

$$\prod_{k=1}^{n_1}\prod_{\ell=1}^{n_2}\left(1 + e^{\beta} \exp\left(-\mathbf{i}\pi\frac{2k + a}{n_1}\right) + e^{\alpha} \exp\left(\mathbf{i}\pi\frac{2\ell + b}{n_2}\right)\right). \tag{4.1}$$

Using this claim, we find the asymptotic behavior of the (weighted) number of possible matchings in the large-scale limit.

Theorem 4.3   *For the weights of edges $\left\{1, e^{\alpha}, e^{\beta}\right\}$, as in Figure 3.2, we have*

$$\lim_{n_1, n_2 \to \infty} \frac{\ln(Z(n_1, n_2))}{n_1 n_2} = \oiint_{|z| = |w| = 1} \ln(|1 + e^{\alpha}z + e^{\beta}w|)\frac{dw}{2\pi\mathbf{i}w}\frac{dz}{2\pi\mathbf{i}z}. \tag{4.2}$$

Remark 4.4   The quantity $\frac{\ln(Z(n_1, n_2))}{n_1 n_2}$ is known as the *free energy* per site.

Exercise 4.5   Show that the value of the double integral remains unchanged if we remove the $|\cdot|$ in (4.2).

Proof of Theorem 4.3 We first consider the case when $e^{\alpha} + e^{\beta} < 1$. In this case, the integral has no singularities, and we conclude that

$$\lim_{n_1,n_2 \to \infty} \frac{\ln(|\det K_{a,b}|)}{n_1 n_2} = \oiint_{|z|=|w|=1} \ln(|1 + e^{\alpha}z + e^{\beta}w|) \frac{dw}{2\pi i w} \frac{dz}{2\pi i z}$$

because the left-hand side is essentially a Riemann sum for the quantity on the right. When using this limit for the asymptotics of $Z(n_1, n_2)$, the only possible issue is that upon taking the $\pm$ for each of a, b, there may be a magic cancellation in $Z(n_1, n_2)$, resulting in it being lower order. To see that no such cancellation occurs, note that

$$\max_{a,b} |\det K_{a,b}| \leq Z(n_1, n_2) \leq 2 \max_{a,b} |\det K_{a,b}|.$$

The second inequality is immediate and follows from the formula for $Z(n_1, n_2)$. The first one follows from noting that $|\det K_{a,b}|$ counts the sets of weighted matchings with $\pm$ values, whereas $Z(n_1, n_2)$ counts the weighted matchings unsigned. The result then follows in this case.

In the case when $e^{\alpha} + e^{\beta} \geq 1$, we may have singularities; we argue that these singularities do not contribute enough to the integral to matter. The easiest way to see this is to consider a triangle with side lengths $1, e^{\alpha}, e^{\beta}$. This gives two angles θ, ϕ, which are the critical arguments (angles) of z and w for singularity to occur. However, note that $\pi i(\frac{2k+a}{n_1})$ cannot simultaneously be close for some k to the critical angle ϕ for both $a = 0$ and $a = 1$. Doing similarly for θ, it follows that there exists a choice of a, b for which one can bound the impact of the singularity on the convergence of the Riemann sum to the corresponding integral, and the result follows. (In particular, one notices that a roughly constant number of partitions in the Riemann sum will have $|1 + e^{\alpha}z + e^{\beta}w|$ close to zero, and foregoing guarantees that you are at least $\geq \frac{1}{\max(n_1,n_2)}$ away from zero, so the convergence still holds.) □

4.2 Densities of Three Types of Lozenges

We can use the free energy computation of the previous section to derive probabilistic information on the properties of the directions of the tiles in a random tiling (or perfect matching). Recall that the probability distribution in question assigns to each tiling (or perfect matching) T the probability

$$\mathbb{P}(T) = \frac{1}{Z(n_1, n_2)} \prod_{\text{Lozenges in } T} \mathbf{w}(\text{lozenge}).$$

Theorem 4.6   *For lozenges ◊ of weight e^α, we have*

$$\lim_{n_1,n_2\to\infty} \mathbb{P}(given\ lozenge\ ◊\ is\ in\ tiling) = \oiint_{|z|=|w|=1} \frac{e^\alpha z}{1+e^\alpha z+e^\beta w}\frac{dw}{2\pi i w}\frac{dz}{2\pi i z}.$$
(4.3)

Proof   Using the shorthand # to denote the number of objects of a particular type in a tiling, we have

$$Z(n_1,n_2) = \sum_{\text{Tilings}} \exp(\alpha\cdot\#◊ + \beta\cdot\#◇),$$

and therefore

$$\frac{\partial}{\partial\alpha}\ln(Z(n_1,n_2)) = \frac{\sum_{\text{Tilings}} \#◊\cdot\exp(\alpha\cdot\#◊+\beta\cdot\#◇)}{\sum_{\text{Tilings}} \exp(\alpha\cdot\#◊+\beta\cdot\#◇)}$$

$$= \mathbb{E}[\#◊]$$

$$= n_1 n_2\, \mathbb{P}(given\ lozenge\ ◊\ is\ in\ tiling).$$

Therefore, using Theorem 3.10, we have

$\mathbb{P}(given\ lozenge\ ◊\ is\ in\ tiling)$

$$= \lim_{n_1,n_2\to\infty} \frac{1}{n_1 n_2}\frac{\partial}{\partial\alpha}\ln(Z(n_1,n_2))$$

$$= \lim_{n_1,n_2\to\infty} \frac{\frac{1}{n_1 n_2}\frac{\partial}{\partial\alpha}\sum_{a,b=0,1} \pm\det K_{a,b}}{\sum_{a,b=0,1} \pm\det K_{a,b}}$$

$$= \lim_{n_1,n_2\to\infty} \sum_{a,b=0,1} \frac{\frac{\partial}{\partial\alpha}\det K_{a,b}}{n_1 n_2 \det K_{a,b}}\cdot\frac{\pm\det K_{a,b}}{\sum_{a,b=0,1} \pm\det K_{a,b}}.$$
(4.4)

Ignoring the possible issues arising from the singularities of the integral, we can repeat the argument of Theorem 4.3, differentiating with respect to α at each step, to conclude that

$$\lim_{n_1,n_2\to\infty} \frac{\frac{\partial}{\partial\alpha}\det K_{a,b}}{n_1 n_2 \det K_{a,b}} = \oiint_{|z|=|w|=1} \frac{e^\alpha z}{1+e^\alpha z+e^\beta w}\frac{dw}{2\pi i w}\frac{dz}{2\pi i z},$$
(4.5)

where the last expression can be obtained by differentiating the right-hand side of (4.2). On the other hand, the four numbers

$$\frac{\pm\det K_{a,b}}{\sum_{a,b=0,1} \pm\det K_{a,b}}$$

have absolute values bounded by 1, and these numbers sum up to 1. Hence, (4.4) implies the validity of (4.3).

As in the previous theorem, we have to address cases of singularities in the integral in order to complete the proof of the statement. To simplify our analysis, we only deal with a subsequence of n_1, n_2 tending to infinity, along which (4.5) is easy to prove. The ultimate statement certainly holds for the full limit as well.

We use the angles θ and ϕ from the proof of Theorem 4.3. Define $W_\theta \subseteq \mathbb{N}$ as

$$W_\theta = \left\{ n : \min_{1 \leq j \leq n} \left| \theta - \frac{\pi j}{n} \right| > \frac{1}{n^{\frac{3}{2}}} \right\}.$$

Define W_ϕ similarly. Note here that θ and ϕ are the critical angles coming from the triangle formed by $1, e^\alpha, e^\beta$.

Lemma 4.7 *For a large enough n, either n or $n + 2$ is in W_θ.*

Proof Suppose that $|\theta - \frac{\pi j}{n}| \leq \frac{1}{n^{\frac{3}{2}}}$ and $|\theta - \frac{\pi j'}{n+2}| \leq \frac{1}{(n+2)^{\frac{3}{2}}}$. Note that for an n that is sufficiently large, this implies that $j' \in \{j, j+1, j+2\}$. However, note that for such j and j', it follows that

$$\left| \frac{\pi j}{n} - \frac{\pi j'}{n+2} \right| = O\left(\frac{1}{n} \right),$$

and thus for a sufficiently large n, we have a contradiction. $\qquad\square$

Now take $n_1 \in W_\phi$ and $n_2 \in W_\theta$. Then note that the denominator in (4.5) (which we get when differentiating in α the logarithm of (4.1)) satisfies

$$\left| 1 + e^\beta \exp\left(-\mathbf{i} \cdot \pi \frac{2k+a}{n_1} \right) + e^\alpha \exp\left(\mathbf{i} \cdot \pi \frac{2\ell+b}{n_2} \right) \right| \geq C \cdot \left(\frac{1}{\min(n_1, n_2)} \right)$$

for all but a set of not more than four critical (k, ℓ) pairs. For these four critical pairs, we have that

$$\left| 1 + e^\beta \exp\left(-\mathbf{i} \cdot \pi \frac{2k+a}{n_1} \right) + e^\alpha \exp\left(\mathbf{i} \cdot \pi \frac{2\ell+b}{n_2} \right) \right| \geq C \cdot \left(\frac{1}{\min(n_1, n_2)^{\frac{3}{2}}} \right)$$

because we have taken $n_1 \in W_\phi$ and $n_2 \in W_\theta$. This bound guarantees that the appropriate Riemann sum converges to the integral expression shown in (4.5). $\qquad\square$

Remark 4.8 Using the same proof as for formula (4.3), we can show that

$$\lim_{n_1, n_2 \to \infty} \mathbb{P}(\text{given lozenge } \lozenge \text{ is in tiling}) = \oiint_{|z|=|w|=1} \frac{1}{1 + e^\alpha z + e^\beta w} \frac{dw}{2\pi \mathbf{i} w} \frac{dz}{2\pi \mathbf{i} z},$$

and

$$\lim_{n_1, n_2 \to \infty} \mathbb{P}(\text{given lozenge } \diamond \text{ is in tiling}) = \oiint_{|z|=|w|=1} \frac{e^\beta w}{1 + e^\alpha z + e^\beta w} \frac{dw}{2\pi \mathbf{i} w} \frac{dz}{2\pi \mathbf{i} z}.$$

We introduce the notation for these probabilities as $p_{\diagdown}$, $p_{\diamond}$ and the one in the theorem statement as $p_{\diagup}$.

Definition 4.9   The (asymptotic) slope of tilings on the torus is $(p_{\diamond}, p_{\diagdown}, p_{\diagup})$.

4.3 Asymptotics of Correlation Functions

The computation of Theorem 4.6 can be extended to arbitrary correlation functions. The extension relies on the following version of Theorem 3.5:

Theorem 4.10   *Take a random lozenge tiling on the torus with arbitrary weights. Write n lozenges as $\ell_i = (w_i, b_i)$ for $i = 1, \ldots, n$. Then*

$$\mathbb{P}(\ell_1, \ldots, \ell_N \in tiling)$$

$$= \frac{\sum_{a,b=0,1} \pm \det K_{ab} \prod_{i=1}^{n} K_{ab}(w_i, b_i) \det_{1 \le i,j \le n} K_{ab}^{-1}(b_i, w_j)}{\sum_{a,b=0,1} \pm \det K_{ab}}.$$

The proof here is nearly identical to the simply connected case, and we omit it. Then, using Theorem 4.10 as an input, we can calculate the asymptotic nth correlation function.

Theorem 4.11   *The nth correlation function in the $n_1, n_2 \to \infty$ limit is*

$$\lim_{n_1,n_2 \to \infty} \mathbb{P}((x_1, y_1, \tilde{x}_1, \tilde{y}_1), \ldots, (x_n, y_n, \tilde{x}_n, \tilde{y}_n) \in tiling)$$

$$= \prod_{i=1}^{n} K_{00}(x_i, y_i, \tilde{x}_i, \tilde{y}_i) \det_{1 \le i,j, \le n} (\tilde{K}^{\alpha,\beta}[\tilde{x}_i - x_j, \tilde{y}_i - y_j]), \quad (4.6)$$

where

$$\tilde{K}^{\alpha,\beta}(x, y) = \oiint_{|z|=|w|=1} \frac{w^x z^{-y}}{1 + e^{\alpha} z + e^{\beta} w} \frac{dw}{2\pi i w} \frac{dz}{2\pi i z}. \quad (4.7)$$

Note that in this theorem, (x_i, y_i) corresponds to the coordinates of the white vertices, and $(\tilde{x}_i, \tilde{y}_i)$ corresponds to the coordinates of the black vertices. The coordinate system here is as in Figure 3.2. Therefore, $(x_i, y_i, \tilde{x}_i, \tilde{y}_i)$ is simply referring to a particular lozenge.

Sketch of the Proof of Theorem 4.11   Here is the plan of the proof:

- We know both the eigenvalues and eigenvectors for the matrix K, and this immediately gives the same for the inverse matrix K^{-1}.
- The key step in analyzing the asymptotic of the expression in Theorem 4.10 is to write down the elements of the matrix inverse as a sum of the coordinates of eigenvectors multiplied by the inverses of the eigenvalues of K. As before, this gives a Riemann sum approximation to the asymptotic formula shown

in (4.7). Note that one can handle singularities in a manner similar to that of Theorem 4.6 in order to get convergence along a subsequence.

- In more detail, the *normalized* eigenvectors of K_{ab} are

$$(x, y) \mapsto \frac{1}{\sqrt{n_1 n_2}} \exp\left(\mathbf{i} \cdot \pi \tfrac{2k+a}{n_1} x + \mathbf{i} \cdot \pi \tfrac{2\ell+b}{n_2} y\right),$$

and the corresponding eigenvalues are

$$1 + e^\beta \exp(-\mathbf{i} \tfrac{2\pi k+a}{n_1}) + e^\alpha \exp(\mathbf{i} \tfrac{2\pi \ell+b}{n_2}).$$

Now note that K_{ab}^{-1} can be written as $Q^{-1} \Lambda^{-1} Q$, where the columns of Q are simply the eigenvectors of K_{ab}, and Λ is the corresponding diagonal matrix of eigenvalues. Note here that $Q^{-1} = Q^*$ because Q corresponds to a discrete two-dimensional Fourier transform. Substituting, it follows that

$$K_{ab}^{-1}(x, y; 0, 0) = \frac{1}{n_1 n_2} \sum_{k=1}^{n_1} \sum_{\ell=1}^{n_2} \frac{\exp(\mathbf{i} \cdot \pi \tfrac{2k+a}{n_1} x + \mathbf{i} \cdot \pi \tfrac{2\ell+b}{n_2} y)}{1 + e^\beta \exp(-\mathbf{i} \tfrac{2\pi k+a}{n_1}) + e^\alpha \exp(\mathbf{i} \tfrac{2\pi \ell+b}{n_2})},$$
(4.8)

and the expression in (4.7) appears as a limit of the Riemann sum of the appropriate integral. (Note that $K_{ab}^{-1}(x, y; 0, 0) = K_{ab}^{-1}((x + \hat{x}, y + \hat{y}; \hat{x}, \hat{y}))$ because the torus is shift invariant.) □

Remark 4.12 The asymptotic of the free energy per site on the $n_1 \times n_2$ torus was investigated for numerous models of statistical mechanics going well beyond weighted tilings, which were investigated in the largest generality in Kenyon et al. (2006). A widely used approach is to treat the torus as a sequence of n_2 strips of size $n_1 \times 1$. One then introduces the *transfer matrix T* encoding the transition from a strip to the adjacent one and computes the free energy as Trace(T^{n_2}). Hence, the computation reduces to the study of the eigenvalues of T, and systematic methods for the evaluation of these eigenvalues are discussed in Baxter (2007). The answer is typically presented in terms of solutions to a system of algebraic equations known as "Bethe equations." Finding analytic expressions for the solutions of these equations (in the limit of the large torus) is a hard task, which is still mathematically unresolved in many situations. From this perspective, tilings represent a specific "nice" case, in which relatively simple formulas of Theorems 4.3 and 4.11 are available.

5

Lecture 5: Monotonicity and Concentration for Tilings

In the last lecture we studied tilings on the torus, and now we go back to planar domains. The setup for this lecture is slightly different from before. We fix a boundary of our domain and values of a height function on the boundary, and we consider all height functions extending the given one. This is equivalent to studying random tilings because we have a correspondence between height functions (up to shifting by a constant) and tilings. The advantage of this setup is that we can compare height functions much easier than comparing tilings directly. The material of this lecture is based on Cohn et al. (1996, 2001). Our aim here is to establish monotonicity and concentration properties for random tilings, which will eventually lead to the Law of Large Numbers for tilings of large domains.

5.1 Monotonicity

We start with a domain $\mathcal{R}$ with boundary $\partial\mathcal{R}$ on the triangular grid. Recall the three positive directions we defined in Lecture 1: $(0, 1)$, $(-\frac{\sqrt{3}}{2}, -\frac{1}{2})$, and $(\frac{\sqrt{3}}{2}, -\frac{1}{2})$. Also recall that along an edge in the positive direction, the height function changes by $+1$ if the edge is an edge of a lozenge, and the height function changes by -2 if the edge crosses a lozenge. Fix a height function h on $\partial\mathcal{R}$. Note that if h changes by -2 on the boundary, then a lozenge is half outside $\mathcal{R}$. Let us emphasize that this is a slightly different situation compared with the previous lectures because the lozenges are allowed to stick out of $\mathcal{R}$; in other words, we are now considering tilings not of $\mathcal{R}$ but of another domain obtained by adding to $\mathcal{R}$ some triangles along the boundary $\partial\mathcal{R}$.

Let us consider a height function H on $\mathcal{R}$ chosen uniformly randomly from all height functions extending h. In the following, $\partial\mathcal{R} \subseteq \mathcal{R}$ can be any set of vertices such that every vertex in $\mathcal{R} \backslash \partial\mathcal{R}$ is adjacent only to vertices of $\mathcal{R}$.

Proposition 5.1 *Let $\mathcal{R}$ and $\partial\mathcal{R}$ be as previously stated. Let h and g be two height functions on $\partial\mathcal{R}$. Assume that $h \equiv g$ (mod 3), $h \geq g$, and that h and g can be extended to height functions on $\mathcal{R}$. Let H (respectively, G) be a height function on $\mathcal{R}$ chosen uniformly randomly from all height functions extending h (respectively, g). Then we can couple (that is, define on the same probability space) H and G in such a way that $H \geq G$ almost surely.*

Remark 5.2   We require that $h \equiv g$ (mod 3) because all height functions are the same modulo 3 up to shifting by a global constant.

Proof of Proposition 5.1   We fix $\mathcal{R}$ and perform induction on $|\mathcal{R} \backslash \partial\mathcal{R}|$. The base case is $\mathcal{R} = \partial\mathcal{R}$, where there is nothing to prove.

Suppose $\partial\mathcal{R} \subsetneq \mathcal{R}$. Choose a vertex $v \in \partial\mathcal{R}$ adjacent to $w \in \mathcal{R} \backslash \partial\mathcal{R}$, such that $h(v) > g(v)$. (If no such v can be chosen, then we can couple H and G in such a way that $H|_{\mathcal{R} \backslash \partial\mathcal{R}} = G|_{\mathcal{R} \backslash \partial\mathcal{R}}$, and there is nothing to prove.) Let us sample H and G on $\mathcal{R} \backslash \partial\mathcal{R}$ in two steps: First, we sample $H(w)$ and $G(w)$, and then we sample the rest.

Because $h(v) > g(v)$ and $h(v) \equiv g(v)$ (mod 3), we have $h(v) \geq g(v) + 3$. Set $x = h(v) - 2$ and $y = g(v) - 2$. The definition of the height function then implies $H(w) \in \{x, x + 3\}$ and $G(w) \in \{y, y + 3\}$. Because $x \geq y + 3$, we can couple $H(w)$ and $G(w)$ so that $H(w) \geq G(w)$.

It remains to couple the values of H and G on $\mathcal{R} \backslash (\partial\mathcal{R} \cup \{w\})$ by the induction hypothesis.   □

Corollary 5.3   *In the setting of Proposition 5.1, for any $w \in \mathcal{R}$, we have $\mathbb{E}\,H(w) \geq \mathbb{E}\,G(w)$.*

Remark 5.4   The proof of Proposition 5.1 extends from the uniform distribution on height functions to more general ones. The only feature of the distribution that we used is the possibility of sequential (Markovian) sampling of the values of the height function and the fact that the set of admissible values for the height function at a point can be reconstructed by the value at a single neighbor. Here is one possible generalization:[1]

Exercise 5.5   Prove an analogue of Proposition 5.1 for the measure on lozenge tilings with weights $\frac{1}{Z} \prod \mathbf{w}(\text{lozenges})$, where $\mathbf{w}(\cdot)$ is a positive function of the type of lozenge and its position; the product goes over all lozenges in a tiling, and Z is a normalizing constant, making the total mass of measure equal to 1.

Similar generalizations to nonuniform distributions are possible for the further results in this lecture.

---

[1]   Another generalization of tilings is the six-vertex model. The monotonicity property extends only to some particular cases of the latter but not to the model with arbitrary parameters. On the level of the proofs, the difficulty arises because the weights of the six-vertex model are determined by the local configurations in squares (x, y), $(x, y + 1)$, $(x + 1, y)$, $(x + 1, y + 1)$ and hence it is not enough to know $h(x, y + 1)$ and $h(x + 1, y)$ in order to sample $h(x + 1, y + 1)$; we also need $h(x, y)$.

Proposition 5.6   *Let $\mathcal{R}_1$ and $\mathcal{R}_2$ be two domains with boundaries $\partial \mathcal{R}_1$ and $\partial \mathcal{R}_2$, respectively. Let h be a height function on $\partial \mathcal{R}_1$, and let g be a height function on $\partial \mathcal{R}_2$ such that $h|_{\partial \mathcal{R}_1 \cap \partial \mathcal{R}_2} \equiv g|_{\partial \mathcal{R}_1 \cap \partial \mathcal{R}_2}$ (mod 3). Suppose that each vertex v of $\partial \mathcal{R}_1$ is within distance Δ_1 from a vertex w in $\partial \mathcal{R}_2$, and vice versa, and for each such pair of vertices, we have $|h(v) - g(w)| \leq \Delta_2$. Then there exists an absolute constant $C > 0$ such that*

$$|\mathbb{E}\, H(x) - \mathbb{E}\, G(x)| \leq C(\Delta_1 + \Delta_2) \text{ for all } x \in \mathcal{R}_1 \cap \mathcal{R}_2,$$

where H (respectively, G) is chosen uniformly randomly from height functions on $\mathcal{R}_1$ (respectively, $\mathcal{R}_2$) extending h (respectively, g).

Proof   Let $\tilde{h}$ be the minimal extension of h to $\mathcal{R}_1$, and let $\tilde{g}$ be the maximal extension of g to $\mathcal{R}_2$. (They exist and are unique because the point-wise maximum or minimum of two height functions is again a height function.) For any vertex $v \in \partial(\mathcal{R}_1 \cap \mathcal{R}_2) \subseteq \partial \mathcal{R}_1 \cup \partial \mathcal{R}_2$, the conditions imply that $\tilde{h}(v) \geq \tilde{g}(v) - C(\Delta_1 + \Delta_2)$ for some absolute constant $C > 0$. Hence, the same inequality is true for any extension of h to $\mathcal{R}_1$ and g to $\mathcal{R}_2$. By Corollary 5.3, we get $\mathbb{E}\, H(x) \geq \mathbb{E}\, G(x) - C(\Delta_1 + \Delta_2)$ for all $x \in \mathcal{R}_1 \cap \mathcal{R}_2$. By swapping $\mathcal{R}_1$ and $\mathcal{R}_2$, we get $\mathbb{E}\, G(x) \geq \mathbb{E}\, H(x) - C(\Delta_1 + \Delta_2)$.   □

Remark 5.7   In the proofs of this section, we silently assumed that the extensions of the height functions from the boundaries exist. As long as this is true, we do not need to assume that the domains are simply connected.

5.2 Concentration

We proceed to the next question: How close is a random height function to its expectation?

Theorem 5.8   *Let $\mathcal{R}$ be a connected domain with boundary $\partial \mathcal{R}$. Let h be a height function on $\partial \mathcal{R}$. Let H be a uniformly random extension of h to $\mathcal{R}$. Suppose that $w, v \in \mathcal{R}$ are linked by a path of length m. Then*

$$\mathbb{E}(H(w) - \mathbb{E}[H(w)|H(v)])^2 \leq 9m,$$

and for each $c > 0$,

$$\mathbb{P}\left(|H(w) - \mathbb{E}[H(w)|H(v)]| > c\sqrt{m}\right) \leq 2\exp\left(-\frac{c^2}{18}\right). \tag{5.1}$$

Remark 5.9   If we choose as v a point for which the value of $H(v)$ is deterministically known (e.g., it can be a point on the boundary $\partial\mathcal{R}$), then expectations become unconditional, and we get the following:

$$\mathbb{E}(H(w) - \mathbb{E}[H(w)])^2 \le 9m,$$

and for each $c > 0$,

$$\mathbb{P}\left(|H(w) - \mathbb{E}[H(w)]| > c\sqrt{m}\right) \le 2\exp\left(-\frac{c^2}{18}\right). \tag{5.2}$$

Remark 5.10   If the linear size of the domain is proportional to L, then so is the maximum (over w) possible value of m. Hence, (5.2) implies that as $L \to \infty$, the normalized height function $\frac{H(Lx, Ly)}{L}$ concentrates around its expectation. We will soon prove that it converges to a limit, thus showing the existence of a *limit shape* for tilings.

Remark 5.11   It is conjectured that the variance should be of order $O(\ln m)$ rather than $O(m)$, which we prove. As of 2021, the tight bound had been proven only for some specific classes of domains.

Remark 5.12   Similar to Exercise 5.5, the result extends to the measures on tilings with weight $\frac{1}{Z}\prod \mathbf{w}$(lozenges). The proof remains the same.

Proof of Theorem 5.8   Let $x_0 = v, x_1, \ldots, x_m = w$ be a path connecting v and w. Let M_k be the conditional expectation of $H(w)$ given the values of $H(x_0)$, $\ldots, H(x_k)$:

$$M_k = \mathbb{E}[H(w) \mid H(x_0), H(x_1), \ldots, H(x_k)].$$

In particular, $M_0 = \mathbb{E}[H(w)|H(v)]$, and $M_m = H(w)$. The tower property of conditional expectations implies that $M_0, \ldots, M_m$ is a martingale.

For fixed values of $H(x_0), \ldots, H(x_k)$, the random variable $H(x_{k+1})$ takes, at most, two distinct values. Also, if $H(x_{k+1})$ takes two values $a < b$, then we must have $b - a = 3$. By Proposition 5.1, we have

$$\mathbb{E}[H(w) \mid H(x_0), H(x_1), \ldots, H(x_k); H(x_{k+1}) = a]$$
$$\le \mathbb{E}[H(w) \mid H(x_0), H(x_1), \ldots, H(x_k); H(x_{k+1}) = b]$$
$$\le \mathbb{E}[H(w) + 3 \mid H(x_0), H(x_1), \ldots, H(x_k); H(x_{k+1}) = a].$$

(We get the second inequality by looking at the height function shifted up by 3 everywhere.) Therefore, $\mathbb{E}[H(w)|H(x_0), H(x_1), \ldots, H(x_k)]$ and $\mathbb{E}[H(w)| H(x_0), H(x_1), \ldots, H(x_{k+1})]$ differ by 3 at most. Hence

$$\mathbb{E}[(M_m - M_0)^2] = \mathbb{E}\left[\left(\sum_{1 \leq k \leq m} (M_k - M_{k-1})\right)^2\right]$$

$$= \sum_{1 \leq k,l \leq m} \mathbb{E}[(M_k - M_{k-1})(M_l - M_{l-1})]$$

$$= \sum_{1 \leq k \leq m} \mathbb{E}[(M_k - M_{k-1})^2] \leq 9m.$$

The third step is because $\mathbb{E}[(M_k - M_{k-1})(M_l - M_{l-1})] = 0$ for $k \neq l$, by the martingale property (taking first the conditional expectation with respect to $H(x_0), H(x_1), \ldots, H(x_{\min(k,l)}))$.

The concentration inequality (5.1) follows from Azuma's inequality applied to the martingale M_k; see Lemma 5.13. $\qquad\square$

Lemma 5.13 (Azuma's Inequality; Azuma, 1967) *Let $X_0, \ldots, X_N$ be a martingale with $|X_k - X_{k-1}| \leq b_k$ for all $1 \leq k \leq N$. Then, for each $t > 0$,*

$$\mathbb{P}(|X_N - X_0| \geq t) \leq 2 \exp\left(-\frac{t^2}{2 \sum_{k=1}^{N} b_k^2}\right).$$

5.3 Limit Shape

The results of the previous two sections say that random height functions concentrate around their expectations, and the expectations continuously depend on the boundary conditions. The limiting profile, which we thus observe as the size of the domain L tends to infinity, is called the "limit shape." So far, we do not have any description for it, and our next aim is to develop such a description. Let us give a preview of the statement that we will be proving. The details are presented in the several following lectures.

We can encode the gradient to (asymptotic) height function via proportions of three types of lozenges $\diamond, \varnothing, \varnothing$. We then define (minus) surface tension to be

$$S(\nabla h) = S(p_\diamond, p_\varnothing, p_\varnothing) = \frac{1}{\pi}(L(\pi p_\diamond) + L(\pi p_\varnothing) + L(\pi p_\varnothing)),$$

where

$$L(\theta) = -\int_0^\theta \ln(2 \sin t)\, dt$$

is the *Lobachevsky function.*

Exercise 5.14 Assume that $p_\diamond$, $p_\square$, $p_\square$ are nonnegative and satisfy $p_\diamond + p_\square + p_\square = 1$. Show that $S(p_\diamond, p_\square, p_\square)$ vanishes whenever one of the arguments vanishes, and $S(p_\diamond, p_\square, p_\square) > 0$ otherwise.

Theorem 5.15 (Variational Principle; Cohn et al., 2001) *Consider a domain $\mathcal{R}^*$ with piece-wise smooth boundary $\partial\mathcal{R}^*$. Let h_b be a real function on $\partial\mathcal{R}^*$. Take a sequence of tileable lattice domains $\mathcal{R}_L$ with boundaries $\partial\mathcal{R}_L$, with scale depending linearly on $L \to \infty$. Fix a height function h_L on $\partial\mathcal{R}_L$. Suppose that $\frac{\partial\mathcal{R}_L}{L} \to \partial\mathcal{R}^*$, and $\frac{h_L}{L}$ converges to h_b in sup norm.[2] Then the height function H_L of a uniformly random tiling of $\mathcal{R}_L$ converges in probability in sup norm. That is, $\frac{H(Lx, Ly)}{L} \to h^{\max}(x, y)$. Furthermore, $h^{\max}$ is the unique maximizer of the integral functional*

$$\iint_{\mathcal{R}^*} S(\nabla h)\, dx\, dy, \qquad h|_{\partial\mathcal{R}^*} = h_b.$$

Simultaneously, we have

$$\frac{1}{L^2} \ln(\text{number of tilings of } \mathcal{R}_L) \to \iint_{\mathcal{R}^*} S(\nabla h^{\max})\, dx\, dy. \qquad (5.3)$$

The proof of Theorem 5.15 is given at the end of Lecture 8 after we do some preparatory work in the next two lectures.

Historically, Theorem 5.15 first appeared with a full proof in the mathematical literature in Cohn et al. (2001). Yet some of its ingredients were implicitly known in theoretical physics earlier; see Nienhuis et al. (1984), Destainville et al. (1997), and Höffe (1997).

---

[2] Because the domain of definition of the function $\frac{\partial h_L}{L}$ depends on L, one should be careful in formalizing the sup-norm convergence. One way is to demand that for each $\varepsilon > 0$ there exists δ and L_0, such that for all $L > L_0$ and all pairs $(x, y) \in \frac{\partial\mathcal{R}_L}{L}$, $(x^*, y^*) \in \partial\mathcal{R}^*$ satisfying $|x - x^*| + |y - y^*| < \delta$, we have $\left|\frac{1}{L} h_L(Lx, Ly) - h_b(x^*, y^*)\right| < \varepsilon$.

6

Lecture 6: Slope and Free Energy

In order to prove Theorem 5.15, we would like to asymptotically count the number of tilings for generic domains. The answer depends on the *slope* of the limit shape for random tilings, so we start with the fixed-slope situation, where our computations for the torus are helpful.

6.1 Slope in a Random Weighted Tiling

We work in the (α, β)-weighted setting, as in Lecture 4. Lozenges of type $\diamond$ have weight e^β, of type $\oslash$ have weight e^α, and of type $\between$ have weight 1, where $\alpha, \beta \in \mathbb{R}$. In Theorem 4.6 and Remark 4.8 we computed the asymptotic slope of random tilings on the torus:

$$p_\diamond = \oiint_{|z|=|w|=1} \frac{e^\beta w}{1 + e^\alpha z + e^\beta w} \frac{dz}{2\pi i z} \frac{dw}{2\pi i w},$$

$$p_\oslash = \oiint_{|z|=|w|=1} \frac{e^\alpha z}{1 + e^\alpha z + e^\beta w} \frac{dz}{2\pi i z} \frac{dw}{2\pi i w},$$

$$p_\between = \oiint_{|z|=|w|=1} \frac{1}{1 + e^\alpha z + e^\beta w} \frac{dz}{2\pi i z} \frac{dw}{2\pi i w}.$$

Let us evaluate one of these integrals, say, the second one, $p_\oslash$. Fix z and compute the w-integral

$$\oint_{|w|=1} \frac{e^\alpha z}{1 + e^\alpha z + e^\beta w} \frac{dw}{2\pi i w}.$$

The integrand has two poles, one at $w = 0$ and one at $w = -\frac{1 + e^\alpha z}{e^\beta}$. If both poles are in the contour, then the w-integral has a value of 0 because we can let the contour go to ∞, and the residue at ∞ is 0 because of $O(\frac{1}{w^2})$ decay. If

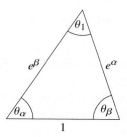

Figure 6.1 Triangle with sides e^α, e^β, 1.

$w = -\frac{1+e^\alpha z}{e^\beta}$ is outside the contour, then the w-integral evaluates to $\frac{e^\alpha z}{1+e^\alpha z}$ by Cauchy's integral formula. Thus, we get

$$p_{\square} = \oint_{|1+e^\alpha z|>e^\beta, |z|=1} \frac{e^\alpha z}{1+e^\alpha z}\frac{dz}{2\pi i z}. \tag{6.1}$$

Suppose there exists a triangle with side lengths e^α, e^β, 1, as in Figure 6.1. Let θ_α and θ_β be the angles opposite to the e^α and e^β edges, respectively. Take $\theta = \pi - \theta_\beta$ so that $|1 + e^\alpha \exp(\mathbf{i}\theta)| = e^\beta$. The integral (6.1) evaluates to

$$p_{\square} = \frac{1}{2\pi i} \ln(1 + e^\alpha z)\Big|_{z=\exp(-i\theta)}^{z=\exp(i\theta)} = \frac{\theta_\alpha}{\pi}.$$

We can compute $p_\diamond$ and $p_\searrow$ similarly and get the following theorem:

Theorem 6.1 *Suppose there exists a triangle with side lengths e^α, e^β, 1, and let its angles be θ_α, θ_β, θ_1, as in Figure 6.1. Then*

$$p_\diamond = \frac{\theta_\beta}{\pi}, \quad p_{\square} = \frac{\theta_\alpha}{\pi}, \quad p_\searrow = \frac{\theta_1}{\pi}.$$

In words, the density of a lozenge is encoded by the angle opposite *to the side of its weight.*

The case when a triangle does not exist is left as an exercise.

Exercise 6.2 1. If $e^\alpha \geq e^\beta + 1$, then $p_{\square} = 1$.
2. If $e^\beta \geq e^\alpha + 1$, then $p_\diamond = 1$.
3. If $1 \geq e^\alpha + e^\beta$, then $p_\searrow = 1$.

Corollary 6.3 *Any slope $(p_\diamond, p_{\square}, p_\searrow)$ with $p_\diamond + p_{\square} + p_\searrow = 1$, $p_\diamond, p_{\square}, p_\searrow > 0$ can be achieved from (α, β)-weighted tilings by choosing α and β appropriately.*

The observation of Corollary 6.3 is the key idea for the Cohn–Kenyon–Propp variational principle.

6.2 Number of Tilings of a Fixed Slope

Now we compute the number of tilings of a torus of a *fixed slope*. Theorem 4.3 computes the asymptotic of the number of tilings as a function of α and β, whereas Theorem 6.1 relates α and β to the slope. So essentially, all we need to do is to combine these two theorems together. We present a way to do it through the Legendre duality. We start from the setting of the finite torus in order to see how this duality arises. (In principle, this is not strictly necessary because we can analyze the formulas of Theorems 4.3 and 6.1 directly.)

Say the torus has dimensions $n_1 \times n_2$. Somewhat abusing the notations, we define the slope of a fixed tiling as

$$\text{slope} = \frac{1}{n_1 n_2}(\#\text{lozenges of types}(\diamondsuit, \lozenge, \lozenge)).$$

(Previously, we thought of the slope only as a limit of this quantity as n_1 and n_2 tend to infinity.)

We have already computed the asymptotics of the partition function Z as a function of α and β in Theorem 4.3. Z is related to the slope by

$$\frac{1}{n_1 n_2} \ln Z = \frac{1}{n_1 n_2} \ln \sum_{\text{slope}} \exp(\alpha \# \lozenge + \beta \# \diamondsuit) \cdot \#\text{tilings of this slope}. \quad (6.2)$$

Claim 6.4 *1. Under the (α, β) measure, the distribution of the slope as a random variable is concentrated as $n_1, n_2 \to \infty$ around $(p_\diamondsuit, p_\lozenge, p_\lozenge)$, which we have computed. This is proved later in this lecture.*

2. The asymptotics of the number of tilings depends continuously on the slope. This is proved in Lecture 8.

Using the claim, we have

$$\frac{1}{n_1 n_2} \ln Z \approx \alpha p_\lozenge + \beta p_\diamondsuit + \frac{1}{n_1 n_2} \ln \#\text{tilings of slope}(p_\diamondsuit, p_\lozenge, p_\lozenge). \quad (6.3)$$

We define the surface tension as

$$\sigma(\text{slope}) = \lim_{n_1, n_2 \to \infty} -\frac{1}{n_1 n_2} \ln \#\text{tilings of such slope}.$$

(This will soon be shown to be the same as $(-1) \cdot S$, with S defined at the end of the previous lecture.)

Let $R(\alpha, \beta)$ be the limit of $\frac{1}{n_1 n_2} \ln Z$ as $n_1, n_2 \to \infty$. Then (6.3) gives

$$R(\alpha, \beta) = \alpha s + \beta t - \sigma(s, t), \quad (6.4)$$

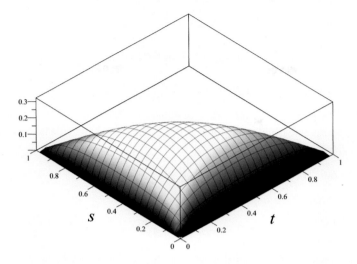

Figure 6.2 The concave function $S(s,t) = -\sigma(s,t)$ appearing in Theorems 5.15 and 6.5.

where (s,t) are asymptotic values for the first two coordinates of the slope: $s = p_\square, t = p_\diamond$. Comparing Theorems 4.3 and 4.6, we also have

$$s = p_\square = \frac{\partial R}{\partial \alpha}, \qquad t = p_\diamond = \frac{\partial R}{\partial \beta}. \qquad (6.5)$$

Here is another point of view on (6.4) and (6.5):

$$R(\alpha, \beta) = \max_{s,t}(\alpha s + \beta t - \sigma(s,t)). \qquad (6.6)$$

The last formula can also be obtained directly: indeed, $(s,t) = (p_\diamond, p_\square)$ is (approximately) the slope with the *largest* probability under the (α, β) measure, and $R(\alpha, \beta)$ is the asymptotic of the logarithm of this probability.

The formulas (6.4), (6.5), (6.6) mean that $R(\alpha, \beta)$ and $\sigma(s,t)$ are *Legendre duals* of each other. Legendre duality is best defined for the convex functions, and in our situation $R(\alpha, \beta)$ is indeed convex in α and β. (This can be checked directly by computing the second derivatives in the formula for $R(\alpha, \beta)$ of Theorem 4.3 and using the Cauchy–Schwarz inequality.)

Because the Legendre duality of R and σ is a symmetric relation, we also have

$$\frac{\partial \sigma}{\partial s} = \alpha, \qquad \frac{\partial \sigma}{\partial t} = \beta.$$

This gives a formula for $\sigma(s,t)$.

Theorem 6.5 *Under the identification $s = p_\emptyset$ and $t = p_\diamond$, we have*

$$\sigma(s, t) = -\frac{1}{\pi}(L(\pi s) + L(\pi t) + L(\pi(1 - s - t))), \qquad (6.7)$$

where $L(\theta) = -\int_0^\theta \ln(2 \sin t)\, dt$ is the Lobachevsky function.

Proof When $s = t = 0$, both sides are equal to 0. Indeed, $\sigma(0, 0) = 0$ because there is exactly one tiling with slope $(0, 0, 1)$. The right-hand side is 0 is because $L(0) = L(\pi) = 0$.

Exercise 6.6 Prove that $L(0) = L(\pi) = 0$.

Let us compare the derivatives:

$$\frac{\partial \mathrm{RHS}}{\partial s} = \ln(2 \sin(\pi s)) - \ln(2 \sin(\pi(1 - t - s))) = \alpha = \frac{\partial \sigma}{\partial s}.$$

The second step is the law of sines applied to the triangle with side lengths e^α, e^β, 1. Similarly, we can show that

$$\frac{\partial \mathrm{RHS}}{\partial t} = \beta = \frac{\partial \sigma}{\partial t}. \qquad \square$$

For the graph of the $\sigma(s, t)$ function, see Figure 6.2.

Proposition 6.7 $\sigma(s, t)$ *is a convex function of (s, t) with $s \geq 0$, $t \geq 0$, $s + t \leq 1$.*

Proof We can check this directly from formula (6.7). This also follows from the fact that $\sigma(s, t)$ is the Legendre dual of $R(\alpha, \beta)$. $\qquad \square$

6.3 Concentration of the Slope

Now let us prove that the slope concentrates as $n_1, n_2 \to \infty$.

Theorem 6.8 *For the (α, β)-weighted random tilings of an $n_1 \times n_2$ torus,*
$$\lim_{n_1, n_2 \to \infty} \frac{|\#\diamond - \mathbb{E}\#\diamond|}{n_1 n_2} = 0 \text{ in probability.}$$

Proof Let us compute the variance:

$$\mathrm{Var}(\#\diamond) = \mathrm{Var}\left(\sum_{(x,y) \in \text{torus}} I(\diamond \text{ at } (x, y)) \right)$$

$$= \sum_{(x,y) \in \text{torus}} \sum_{(x',y') \in \text{torus}} \mathrm{Cov}(I(\diamond \text{ at } (x, y)), I(\diamond \text{ at } (x', y'))).$$

($I(\cdot)$ denotes the indicator function of an event.) If we show that Cov $\to 0$ as $|x - x'|, |y - y'| \to \infty$, then Var $= o(n_1^2 n_2^2)$, and concentration holds by Chebyshev's inequality. We have

$$\text{Cov} = \mathbb{P}(\diamond \text{ at } (x, y) \text{ and at } (x', y')) - \mathbb{P}(\diamond \text{ at } (x, y))\mathbb{P}(\diamond \text{ at } (x', y')).$$

By Theorem 4.11 of Lecture 4, as $n_1, n_2 \to \infty$, the covariance goes to

$$e^{2\beta} \det \begin{pmatrix} \tilde{K}^{\alpha,\beta}(1, 0) & \tilde{K}^{\alpha,\beta}(x - x' + 1, y - y') \\ \tilde{K}^{\alpha,\beta}(x' - x + 1, y' - y) & \tilde{K}^{\alpha,\beta}(1, 0) \end{pmatrix} - e^{2\beta} \tilde{K}^{\alpha,\beta}(1, 0)^2,$$

$$(6.8)$$

where

$$\tilde{K}^{\alpha,\beta}(x, y) = \oiint_{|z|=|w|=1} \frac{w^x z^{-y}}{1 + e^\alpha z + e^\beta w} \frac{dz}{2\pi i z} \frac{dw}{2\pi i w}.$$

The double contour integral is a Fourier coefficient of a reasonably nice function, so it goes to zero as $|x|, |y| \to \infty$. Therefore, (6.8) also goes to zero, as desired. (Note that we use here slightly more than what is proved in Lecture 4. There, $x - y'$ and $y - y'$ are kept constant as $n_1, n_2 \to \infty$. Here, they should grow with n_1, n_2, so formally, we have to deal directly with discrete Fourier coefficients of (4.8), but those still go to 0.) □

6.4 Limit Shape of a Torus

We end this lecture by discussing how a random tiling of the torus looks on the macroscopic scale.

Take a fundamental domain of the torus. Let $H = 0$ at the bottom-left corner, and use paths only *inside* the fundamental domain to define height function $H(x, y)$ for the torus. By translation invariance, we have $\mathbb{E} H(x, y) - \mathbb{E} H(x - 1, y) = a$ and $\mathbb{E} H(x, y) - \mathbb{E} H(x, y - 1) = b$, for all (x, y). Here, (a, b) is a vector that can be treated as a version of the slope. We conclude that $\mathbb{E} H(x, y) = ax + by$.

Hence, using concentration inequality (5.2) from Lecture 5 (see Remark 5.12), as $L \to \infty$, we have

$$\sup_{x,y} \left| \frac{H(Lx, Ly)}{L} - (ax + by) \right| \to 0$$

in probability. In other words, the limit shape for tilings is a plane.

Lecture 7: Maximizers in the Variational Principle

7.1 Review

Recall the definition of the *Lobachevsky function*, given by:

$$L(\theta) = -\int_0^\theta \ln(2\sin t)\, dt.$$

Also recall the definition of the (minus) surface tension associated with a triple $(p_\Diamond, p_\square, p_\searrow)$ of densities given by

$$S(p_\Diamond, p_\square, p_\searrow) = \frac{1}{\pi}\left(L(\pi p_\Diamond) + L(\pi p_\square) + L(\pi p_\searrow)\right).$$

The setup for the identification of the limit shape in the variational principle of Theorem 5.15 is as follows:

- For $L \to \infty$, we consider a sequence of finite domains $\mathcal{R}_L$, whose scale depends linearly on L, and fix a height function h_L on each boundary $\partial\mathcal{R}_L$. We silently assume that h_L is such that it has extensions to height functions on the whole $\mathcal{R}_L$.
- As our target, we take a continuous region $\mathcal{R}^* \subseteq \mathbb{R}^2$ with piece-wise smooth boundary $\partial\mathcal{R}^*$ and let h_b be a real function on this boundary.

We require the boundaries of $\mathcal{R}_L$ to converge to those of the limiting region $\mathcal{R}^*$:

$$\frac{1}{L}\partial\mathcal{R}_L \to \partial\mathcal{R}^*.$$

We do not need to specify the precise notion of convergence here because the limit shapes do not depend on the small perturbations of the boundaries (see Lectures 5 and 8). Moreover, we require that

$$\frac{h_L(Lx, Ly)}{L} \to h_b(x, y) \quad \text{on } \partial\mathcal{R}^*,$$

which means that for each $\varepsilon > 0$, there exists $\delta > 0$ and L_0, such that for all $L > L_0$ and all pairs $(x, y) \in \frac{\partial \mathcal{R}_L}{L}$, $(x^*, y^*) \in \partial \mathcal{R}^*$ satisfying $|x - x^*| + |y - y^*| < \delta$, we have $\left|\frac{1}{L} h_L(Lx, Ly) - h_b(x^*, y^*)\right| < \varepsilon$.

The variational principle asserts (among other things) that a uniformly random height function H_L of $\mathcal{R}_L$ extending h_L on $\partial \mathcal{R}_L$ (= height function of a uniformly random tiling) converges in probability to a function h^* given by

$$h^* = \operatorname{argmax}_{h \in \mathcal{H}} \iint_{\mathcal{R}^*} S(\nabla h) \, dx \, dy. \tag{7.1}$$

In other words, the limit shape should be one that maximizes a certain surface-tension integral.

The goal of this lecture is merely to make (7.1) precise (and subsequent lectures will actually prove the theorem). This means we need to take the following steps:

- We define and motivate the class of functions $\mathcal{H}$ mentioned in maximization problem (7.1).
- We define $S(\nabla h)$ for $h \in \mathcal{H}$ by associating a triple $(p_\diamond, p_\lhd, p_\searrow)$ of local densities to it.
- We prove that there exists a maximizing function h^* across the choice of $h \in \mathcal{H}$.
- We prove that the maximizing function is unique.

Throughout this lecture, we retain $\mathcal{R}^*$ as our target region with piece-wise smooth boundary $\partial \mathcal{R}^*$. We fix the boundary function h_b on $\partial \mathcal{R}^*$ as well.

7.2 The Definition of Surface Tension and Class of Functions

7.2.1 Identifying the Gradient with the Probability Values

Let's consider once again an h on a finite region. We can imagine walking in the three directions indicated in Figure 7.1. Recall that the way our height function h is defined, as we move in the $+y$ direction, we get an increment of either $+1$ or -2 according to whether we cross a lozenge or not. Accordingly, in the standard coordinate system with the x-axis pointing to the right and the y-axis pointing up, we might colloquially write

$$\frac{\partial h}{\partial y} = -2p_\diamond + p_\searrow + p_\lhd = 1 - 3p_\diamond.$$

Similarly, moving in the other directions gives the following:

$$+\frac{\sqrt{3}}{2}\frac{\partial h}{\partial x} - \frac{1}{2}\frac{\partial h}{\partial y} = p_\diamond + p_\searrow - 2p_\lhd = 1 - 3p_\lhd$$

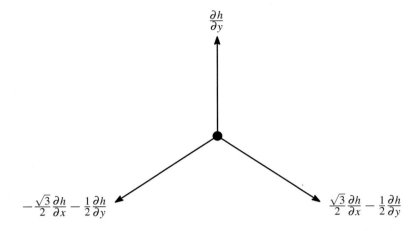

Figure 7.1 Three coordinate directions correspond to three linear combinations of partial derivatives.

$$-\frac{\sqrt{3}}{2}\frac{\partial h}{\partial x} - \frac{1}{2}\frac{\partial h}{\partial y} = p_\diamond - 2p_\searrow + p_\square = 1 - 3p_\searrow.$$

In this way, we can now create a definition that gives a local density triple $(p_\diamond, p_\square, p_\searrow)$ at a point for a *differentiable* function $h\colon \mathcal{R}^* \to \mathbb{R}$.

Definition 7.1 Let $h\colon \mathcal{R}^* \to \mathbb{R}$ be differentiable. For a point w in the interior of $\mathcal{R}^*$, we identify the gradient $(\nabla h)_w$ with the following three local densities:

$$p_\diamond = \frac{1 - \frac{\partial h}{\partial y}}{3},$$

$$p_\square = \frac{1 - \left(-\frac{\sqrt{3}}{2}\frac{\partial h}{\partial x} - \frac{1}{2}\frac{\partial h}{\partial y}\right)}{3},$$

$$p_\searrow = \frac{1 - \left(\frac{\sqrt{3}}{2}\frac{\partial h}{\partial x} - \frac{1}{2}\frac{\partial h}{\partial y}\right)}{3}.$$

Naturally, we have $p_\diamond + p_\searrow + p_\square = 1$. Thus, moving forward, we will just write $S(\nabla h)$ for height functions h with this convention.

7.2.2 The Desired Set of Functions

Next, we define the set of functions $\mathcal{H}$ we wish to consider: they are the functions $h\colon \mathcal{R}^* \to \mathbb{R}$ that satisfy a *Lipschitz* condition so that $p_\diamond, p_\searrow, p_\square$ are in the interval $[0, 1]$.

Definition 7.2 Let $\delta_1, \delta_2, \delta_3$ be unit vectors in the coordinate directions of Figure 7.1. We denote by $\mathcal{H}$ the set of all functions

$$h: \mathcal{R}^* \to \mathbb{R},$$

such that

1. h agrees with h_b (on $\partial \mathcal{R}^*$);
2. h satisfies the Lipschitz condition: for each $i = 1, 2, 3$ and $\ell > 0$

$$-2 \leq \frac{h((x, y) + \ell \delta_i) - h(x, y)}{\ell} \leq 1, \tag{7.2}$$

where (x, y) and $\ell > 0$ can be taken as arbitrary as long as both (x, y) and $(x, y) + \ell \delta_i$ belong to $\mathcal{R}^*$.

Let us recall a general result about the differentiability of Lipshitz functions known as the "Rademacher theorem."

Theorem 7.3 *A Lipschitz function is differentiable almost everywhere.*

Whenever a function $h \in \mathcal{H}$ is differentiable at (x, y), the inequality (7.2) implies that the gradient encoded via Definition 7.1 by local densities $(p_\diamond, p_\triangleright, p_\square)$ satisfies $0 \leq p_\diamond, p_\triangleright, p_\square \leq 1$.

We henceforth take as a standing assumption that $\mathcal{H} \neq \varnothing$; this essentially means that the boundary function h_b should satisfy a similar Lipschitz condition. (Which is automatically true if h_b is obtained as a limit of height functions of tileable regions.)

7.2.3 Goal

We have completed all necessary definitions, and the rest of this lecture is devoted to proving that there is a unique function $h^* \in \mathcal{H}$ that maximizes the value of

$$\iint_{\mathcal{R}^*} S(\nabla h) \, dx \, dy.$$

7.3 Upper Semicontinuity

Definition 7.4 Let M be a metric space. A real-valued function $f: M \to \mathbb{R}$ is *upper semicontinuous* if for every point $p \in M$, $\limsup_{x \to p} f(x) \leq f(p)$.

As the first step in proving the existence/uniqueness result, we prove the following:

Theorem 7.5 *Equip $\mathcal{H}$ with the uniform topology. Then,*

$$h \mapsto \iint S(\nabla h) \, dx \, dy$$

is upper semicontinuous as a function $h \in \mathcal{H}$.

The proof relies on several lemmas.

Lemma 7.6 *Let $h \in \mathcal{H}$. Choose a mesh of equilateral triangles of side length ℓ. Take a function $\widetilde{h}$ agreeing with h at vertices of each triangle and linear on each triangle (hence, piece-wise linear).*

For each $\varepsilon > 0$, if $\ell > 0$ is smaller than $\ell_0 = \ell_0(\varepsilon)$, then for at least $1 - \varepsilon$ fraction of the triangles, the following holds:

- $|\widetilde{h} - h| \le \ell\varepsilon$ *at each point inside the triangle;*
- *for the $(1 - \varepsilon)$ fraction of the points of triangles, ∇h exists, and*
  $\left\| \nabla h - \nabla\widetilde{h} \right\| < \varepsilon.$

Outline of Proof   We mirror the proof of Cohn et al. (2001), Lemma 2.2. We first show that there is a $1 - \varepsilon/3$ fraction of triangles contained strictly within the region satisfying the first condition, then proceed to the second condition. In both parts, we implicitly use the Rademacher theorem, implying that h is differentiable almost everywhere.

- First part: For every point of differentiability w, there exists a small real number $r(w) > 0$ such that

$$|h(w + d) - h(w) - (\nabla h)_w \cdot d| < \frac{1}{10}\varepsilon \cdot |d|$$

  for small displacements $|d| \le r(w)$. If our mesh ℓ is so small that the set

$$\{w \mid r(w) > \ell\} \tag{7.3}$$

  has a measure of at least $(1 - \varepsilon/3) \cdot (\text{Area } \mathcal{R}^*)$, then (ignoring the triangles cut off by the boundary) we can find a point w from set (7.3) in each of the $(1 - \varepsilon/3)$ fractions of the triangles. Thus, the linear approximation property holds for these triangles.
- For the second part, we need to use the *Lebesgue density theorem*: Let A be any Lebesgue measurable set, and define the local density of A at x as

$$\lim_{\epsilon \to 0} \frac{\text{mes}(B_\epsilon(x) \cap A)}{\text{mes}(B_\epsilon(x))},$$

  with $B_\epsilon(x)$ being a ball of radius ϵ around x. Then, the theorem (which we do not prove here) claims that for almost every point $x \in A$, the local density equals 1.

Next, we choose finitely many sets $U_1, U_2, \ldots, U_n$ covering all $\mathcal{R}^*$ and such that whenever $p, q \in U_i$, we have $\left\| (\nabla h)_p - (\nabla h)_q \right\| < \varepsilon/2$ (this is possible because the space of possible gradients ∇h is compact). By applying Lebesgue's density theorem to each U_i, if ℓ is small enough, for point w in all but the $\varepsilon/3$ fraction of $\mathcal{R}^*$, the value of ∇h in all but the $\varepsilon/3$ fraction of the ball of radius ℓ centered at w differs from $(\nabla h)_w$ by not more than $\varepsilon/2$.

If w additionally belongs to the set (7.3) from the first part, then on the triangle to which w belongs, $\nabla \tilde{h}$ differs from $(\nabla h)_w$ by not more than $\varepsilon/2$. Hence, the second property holds. $\qquad\square$

Here is an immediate consequence of Lemma 7.6:

Corollary 7.7 *Retaining the notation of the lemma,*

$$\lim_{\varepsilon \to 0} \left(\iint_{\mathcal{R}^*} S(\nabla h) \, dx \, dy - \iint_{\mathcal{R}^*} S(\nabla \tilde{h}) \, dx \, dy \right) = 0.$$

The next lemma deals with very specific choices of $\mathcal{R}^*$.

Lemma 7.8 *Take as $\mathcal{R}^*$ an equilateral triangle T of side length ℓ. Let $h \in \mathcal{H}$, and choose a linear function $\tilde{h}$ satisfying $\left| h - \tilde{h} \right| < \varepsilon \ell$ on ∂T. Then*

$$\frac{\iint_T S(\nabla h) \, dxdy}{\operatorname{Area} T} \leq \frac{\iint_T S(\nabla \tilde{h}) \, dxdy}{\operatorname{Area} T} + o(1)$$

as $\varepsilon \to 0$.

In other words, we can replace h by its linear approximation if they agree along the border of the domain T. Note that $\tilde{h}$ is linear, rather than piece-wise linear; hence, this lemma is being applied to just one triangle T.

Proof of Lemma 7.8 By concavity of S, we have

$$\frac{\iint_T S(\nabla h) \, dxdy}{\operatorname{Area} T} \leq S\left(\operatorname{Avg}(\nabla h) \right), \tag{7.4}$$

where $\operatorname{Avg}(\nabla h)$ is the average value of ∇h on T. Also, we have

$$\left\| \operatorname{Avg}(\nabla h) - \operatorname{Avg}(\nabla \tilde{h}) \right\| = O(\varepsilon),$$

which can be proven by reducing the integral over T in the definition of the average value to the integral over ∂T using the fundamental theorem of the calculus and then using $\left| h - \tilde{h} \right| < \varepsilon \ell$. Therefore, by the continuity of S,

$$S(\operatorname{Avg}(\nabla h)) = S(\operatorname{Avg}(\nabla \tilde{h})) + o(1).$$

Because $\widetilde{h}$ is linear, its gradient is constant on T, and we can remove the average value in the right-hand side of the last identity. Combining with (7.4), we get the result. $\square$

Proof of Theorem 7.5   Let $h \in \mathcal{H}$. Then, for each $\gamma > 0$ we want to show that there exists $\delta > 0$ such that whenever g and h differ by not more than δ everywhere, we should have

$$\iint_{\mathcal{R}^*} S(\nabla g)\, dx dy \leq \iint_{\mathcal{R}^*} S(\nabla h)\, dx dy + \gamma.$$

Take a piece-wise linear approximation $\widetilde{h}$ for h from Lemma 7.6, and choose $\delta < \varepsilon \ell$. Let us call a triangle T of mesh "good" if two approximation properties of Lemma 7.6 hold for it. Then, as $\varepsilon \to 0$, we have by Lemma 7.8

$$\iint_T S(\nabla g) \leq \iint_T S(\nabla \widetilde{h}) + \mathrm{Area}(T) \cdot o(1),$$

and thus by summing over all the good triangles T, we have

$$\iint_{\substack{\text{good triangles in } \mathcal{R}^*}} S(\nabla g) \leq \iint_{\substack{\text{good triangles in } \mathcal{R}^*}} S(\nabla \widetilde{h}) + \mathrm{Area}(\mathcal{R}^*) \cdot o(1).$$

Because S is bounded, the bad triangles only add another $O(\varepsilon)$. Thus, also using Corollary 7.7, we get

$$\iint_{\mathcal{R}^*} S(\nabla g) \leq \iint_{\mathcal{R}^*} S(\nabla h) + \mathrm{Area}(\mathcal{R}^*) \cdot o(1). \qquad \square$$

7.4 Existence of the Maximizer

The development of the previous section leads to the existence of maximizers.

Corollary 7.9   *There exists a maximizer $h^* \in \mathcal{H}$:*

$$h^* = \mathrm{argmax}_{h \in \mathcal{H}} \iint_{\mathcal{R}^*} S(\nabla h)\, dx dy.$$

Proof   We equip the set $\mathcal{H}$ with the uniform topology. It is then compact by the Arzela–Ascoli theorem.

Let h_n, $n = 1, 2, \ldots$, be a sequence of functions such that the integrals $\iint S(\nabla h_n)\, dx dy$ approach as $n \to \infty$ the supremum $\sup_{h \in \mathcal{H}} \iint S(\nabla h)\, dx dy$. Passing to a convergent subsequence by compactness, we can assume without loss of generality that $h_n \to h^*$, for some $h^* \in \mathcal{H}$. Then, by Theorem 7.5, we have

$$\sup_{h \in \mathcal{H}} \iint S(\nabla h)\, dx dy = \limsup_{n \to \infty} \iint S(\nabla h_n)\, dx dy \leq \iint S(\nabla h^*)\, dx dy. \quad \square$$

On the technical level, the difficulty in the proof of Corollary 7.9 is to pass from uniform convergence of the functions to convergence of their gradients, which is necessary because S is a function of the gradient. This was the main reason for the introduction of the mesh of small equilateral triangles T in the previous section.

7.5 Uniqueness of the Maximizer

We end this lecture with a discussion of the uniqueness of maximizers.

Let h_1 and h_2 be two maximizers (with the same value of the integral of S), and let $h = \frac{h_1+h_2}{2} \in \mathcal{H}$. Then, using the concavity of S, we have

$$\iint S(\nabla h)\, dxdy \geq \frac{1}{2} \left(\iint S(\nabla h_1)\, dxdy + \iint S(\nabla h_2)\, dxdy \right)$$
$$= \iint S(\nabla h_1)\, dxdy. \qquad (7.5)$$

If $S(\bullet)$ were strictly concave, the inequality in (7.5) would have been strict, and this would have been a contradiction with $h_1 \neq h_2$ being two maximizers. Annoyingly, this is not quite true – S is only strictly concave in the region where $p_\diamond, p_\lozenge, p_\varnothing > 0$ (i.e., the interior of the triangle of possible slopes). Nonetheless, we would be done if there are points with $\nabla h_1 \neq \nabla h_2$, with at least one of them strictly inside the triangle of slopes.

Hence, the only remaining case is that all gradients ∇h_1 and ∇h_2 lie on the boundary of the triangle of the slopes. Let us distinguish two kinds of boundary slopes: extreme slopes, such that $(p_\diamond, p_\lozenge, p_\varnothing)$ is either $(1,0,0)$, or $(0,1,0)$, or $(0,0,1)$, and nonextreme boundary slopes where proportions of two types of lozenges are nonzero. Note that if $h_1 \neq h_2$ are two maximizers with all the slopes on the boundary, then passing to $h = \frac{h_1+h_2}{2}$ and using (7.5), if necessary, we would find a maximizer with nonextreme boundary slopes. However, we did not see any such slopes in the simulations shown in Figures 1.3 and 1.5 of Lecture 1. In fact, the following statement explains that nonextreme boundary slopes *never appear* in the limit shapes of tilings, as long as we tile nondegenerate domains.

Proposition 7.10 *If a domain $\mathcal{R}^*$ with h_b on $\partial\mathcal{R}^*$ has at least one height function $h^\circ \in \mathcal{H}$ extending h_b, such that all the slopes ∇h° inside $\mathcal{R}^*$ are not on the boundary of the triangle of possible slopes, then h with nonextreme boundary slopes (i.e., boundary slopes with nonzero proportions of precisely two types of lozenges) cannot be a maximizer of $\iint S$.*

Remark 7.11   For all the domains that we encounter in these lectures, the existence of $h°$ is immediate. For instance, this is the case for tilings of polygonal domains with sides parallel to the six grid directions of the triangular lattice. The only possible "bad" situation that we need to exclude is when the boundary conditions make some part of the tiling frozen (i.e., having only one type of lozenge) for every extension of h_b; in this case, we should simply remove this frozen part from the consideration and study the height functions in the complement.

Outline of the proof of Proposition 7.10   For $\varepsilon > 0$, consider

$$\iint_{\mathcal{R}^*} \left[S(\nabla(\varepsilon h° + (1 - \varepsilon)h)) - S(\nabla h) \right] \, dxdy. \tag{7.6}$$

We would like to study the small ε expansion of this difference by expanding the integrand at each (x, y).

The explicit formula for S implies that the partial derivatives of S are bounded inside the triangle of slopes but grow logarithmically at the boundary (except for the vertices $(1, 0, 0)$, $(0, 1, 0)$, $(0, 0, 1)$); see Figure 6.2.

Hence, the contribution to (7.6) of (x, y) with nonboundary (or extreme boundary) slopes of ∇h is $O(\varepsilon)$. On the other hand, the contribution of (x, y) with nonextreme boundary slopes is of the order $\varepsilon \ln \frac{1}{\varepsilon} \gg \varepsilon$ because the definition of $h°$ implies that $\nabla(\varepsilon h° + (1 - \varepsilon)h)$ is strictly inside the triangle of slopes for any $\varepsilon > 0$. Simultaneously, this contribution is positive for all (x, y) because S has a minimum on the border of the triangle; see Figure 6.2. Hence, (7.6) is positive for small ε, contradicting the maximality of the integral for h.

□

Another proof for the uniqueness of maximizers can be found in Proposition 4.5 of De Silva and Savin (2010).

8

Lecture 8: Proof of the Variational Principle

In this lecture, using the machinery we have set up, we prove the variational principle of Theorem 5.15 following Cohn et al. (2001).

Proposition 8.1 *Let $\mathcal{R}$ be a convex region in the plane, and fix positive reals k and ε. Suppose that $\mathcal{R}$ fits within some $L \times L$ square, where $L > \varepsilon^{-1}$, and suppose it has area $A \geq kL^2$.*

Let h_b be a height function on $\partial \mathcal{R}$ that is within εL of the height function of a fixed plane Π. Then, the quantity

$$\frac{1}{A} \ln(\text{\# of extensions of } h_b \text{ to } \mathcal{R}) \tag{8.1}$$

is independent of the choice of h_b (but depends on Π) up to an additive error of $O(\varepsilon^{1/2} \ln \frac{1}{\varepsilon})$, which might depend on k.

Proof   The proof splits into two cases, depending on the slope of Π.

Case 1: The slope of Π is at least $\varepsilon^{1/2}$ away from any boundary slope; in other words, the proportion of each lozenge corresponding to the slope of Π is at least $\varepsilon^{1/2}$ and at most $1 - \varepsilon^{1/2}$.

Take a height function g_b (on $\partial \mathcal{R}$) also agreeing with the plane Π; we will show that the values of (8.1) for g_b and h_b are within $O(\varepsilon^{1/2} \ln \frac{1}{\varepsilon})$ of each other.

Assume $g_b \geq h_b$; if not, then we can compare both g_b and h_b with $\min(g_b, h_b)$ instead. Let g^* and h^* denote the minimal extension of g_b and the maximal extension of h_b to $\mathcal{R}$, respectively (because the maximum/minimum of two height functions is again a height function, such extensions are uniquely defined). We define two maps that operate on height functions on $\mathcal{R}$. If f is an extension of g_b to $\mathcal{R}$, then we define $H(f)$ to be the extension of h_b to $\mathcal{R}$ given by the point-wise minimum:

$$H(f) = \min(f, h^*).$$

Similarly, if f is an extension of h_b to $\mathcal{R}$, then $G(f)$ is the extension of g_b to $\mathcal{R}$, given by:

$$G(f) = \max(f, g^*).$$

Note that $H(G(f)) = f$ at all points in $\mathcal{R}$ at which the value of f is between g^* and h^* (in that order).

Recall the formula for h^* in terms of h_b from Lecture 1 (for g^*, the formula is similar):

$$h^*(x) = \min_{u \in \partial \mathcal{R}} (d(x, u) + h_b(u)),$$

where d is the distance function within $\mathcal{R}$. Note that because $\mathcal{R}$ is convex, d differs (up to a small error) by a constant factor from the Euclidean distance in the plane. Thus, g^* and h^* differ from the height of Π by at least $\Omega(\varepsilon^{1/2} d(x, \partial \mathcal{R}))$ because the slope of Π is at least $\varepsilon^{1/2}$ away from any boundary slope.

On the other hand, we claim that a typical random extension of h_b to $\mathcal{R}$ is close to Π. Indeed, to see that, we embed $\mathcal{R}$ into a torus and consider a random (α, β)-weighted tiling of the torus, where α and β are chosen so that the limit shape of such a random tiling has the slope Π; see Lecture 6. The concentration of the height function of Lecture 5 applied to the torus means that h_b is close on $\partial \mathcal{R}$ to the height function of the random tiling of the torus. We can then compare the random extension of h_b to a random tiling of the torus conditional on the values of its height function on $\partial \mathcal{R}$. Note that the conditioning removes nonuniformity; that is, the conditional law of (α, β)-weighted tilings on the torus is the same as for the uniformly random tilings.[1] Hence, we can use the arguments of Lecture 5. In particular, Proposition 5.1 (see also Proposition 5.6) implies that with probability tending to 1 (as the size of the domain grows), the height function of a random extension of h_b is close to that of a random tiling of the torus, which is close to Π.

We conclude that with high probability, a random extension of h_b to $\mathcal{R}$ is at distance not more than εL from Π. Thus, we have that for random extension f of h_b, with probability approaching 1, $H(G(f)) = f$ at all points at a distance at

---

[1] In order to see the conditional uniformity, notice that as soon as we know the values of the height function on $\partial \mathcal{R}$, we know the total number of lozenges of each of the three types inside $\mathcal{R}$. Indeed, all the tilings with fixed boundary conditions can be obtained from each other by adding/removing unit cubes, as in the first proof of Theorem 2.1, and each such operation keeps the number of lozenges of each type unchanged. Hence, each lozenge tiling of $\mathcal{R}$ with fixed values of the heights on $\partial \mathcal{R}$ has the same (α, β) weight.

least $\varepsilon^{1/2}(\ln\frac{1}{\varepsilon})L$ from the boundary $\partial\mathcal{R}$. This gives a partial bijection between extensions of h_b and extensions of g_b.

It remains to deal with the points at a distance less than $\varepsilon^{1/2}(\ln\frac{1}{\varepsilon})L$ from the boundary $\partial\mathcal{R}$. The possible values of the height function at these points can be constructed sequentially: when we move from a point to its neighbor, there are two possible choices for the value of the height function. Thus, noting that there are $O(\varepsilon^{1/2}(\ln\frac{1}{\varepsilon})L^2)$ points close to the boundary $\partial\mathcal{R}$, the number of extensions of g and h differ by a multiplicative factor of not more than $2^{O(\varepsilon^{1/2}(\ln\frac{1}{\varepsilon})L^2)}$, as desired.

Case 2: The slope of Π is within $\varepsilon^{1/2}$ of some boundary slope; in other words, the proportion of some type of lozenges, $p_\diamond$, $p_\square$, or $p_\square$, is not more than $\varepsilon^{1/2}$. Without loss of generality, assume that the probability of a horizontal lozenge is $p_\diamond < \varepsilon^{1/2}$. Let us show that in this case, (8.1) is close to 0.

Consider some extension of h_b, and consider all the vertical sections of $\mathcal{R}$ by straight lines. Note that we know the number of horizontal lozenges $\diamond$ on these lines exactly by calculating the change in the value of h_b on each of these lines. This number is $O(\varepsilon^{1/2}L)$ because of h_b being close to Π on $\partial\mathcal{R}$. Thus, the total number of $\diamond$ lozenges inside $\mathcal{R}$ is not more than $O(\varepsilon^{1/2}L^2)$. Note that the positions of $\diamond$ lozenges uniquely determine the rest of the tiling (i.e., the lozenges of the other two types). Thus, the number of tilings is not more than the number of arrangements of $\diamond$ lozenges, which is the following, at most (where c is the constant in the big O):

$$\binom{A}{c\varepsilon^{1/2}L^2} \le \frac{A^{c\varepsilon^{1/2}L^2}}{\left(\frac{c\varepsilon^{1/2}L^2}{e}\right)^{c\varepsilon^{1/2}L^2}} \le \left(\frac{L^2}{\frac{c\varepsilon^{1/2}L^2}{e}}\right)^{c\varepsilon^{1/2}L^2} = \left(\frac{e}{c\varepsilon^{1/2}}\right)^{c\varepsilon^{1/2}L^2}$$

$$= \exp\left(L^2 O(\varepsilon^2\ln\frac{1}{\varepsilon})\right). \qquad\square$$

Corollary 8.2 *In the setting of Proposition 8.1, if $\mathcal{R}$ is an $L\times L$ square, then*

$$\frac{1}{A}\ln(\#\text{ of extensions of }h\text{ to }\mathcal{R}) = S(\text{slope of }\Pi) + O\left(\varepsilon^2\ln\frac{1}{\varepsilon}\right). \quad (8.2)$$

Proof Let us embed the $L\times L$ square into an $L\times L$ torus. If we deal with random tilings of this torus, then the height function on the boundary of our square is now *random*. However, in Section 6.4, we have shown that this height function is asymptotically planar. Hence, using Proposition 8.1, we conclude that the number of tilings of the square is approximately the same as the number of the fixed-slope tilings of the torus.[2] The asymptotic of the latter was computed in Theorem 6.5.[3] □

Proposition 8.3   *The statement (8.2) still holds when $\mathcal{R}$ is an equilateral triangle.*

Proof   Tile $\mathcal{R}$ with small squares so that every square is completely contained within $\mathcal{R}$ (a small portion of $\mathcal{R}$ will be uncovered, but this is negligible, so we ignore it). Fix a (consistent) height function that is within $O(1)$ of the plane on the boundaries of the squares. Note that given this height function on the boundaries of the squares, a tiling of each of the squares yields a valid tiling of $\mathcal{R}$. Therefore,

$$\frac{1}{A}\ln(\text{\# of tilings of } \mathcal{R}) \geq \frac{1}{A}\ln(\text{\# of tilings of all squares})$$

$$= \frac{1}{A}\sum_{\text{squares } s}\ln(\text{\# of tilings of } s)$$

$$\geq \frac{1}{A}\sum_{\text{squares } s}\text{Area}(s)\left(S(\text{slope of } \Pi) + O\left(\varepsilon^2 \ln \tfrac{1}{\varepsilon}\right)\right)$$

$$= S(\text{slope of } \Pi) + O\left(\varepsilon^2 \ln \tfrac{1}{\varepsilon}\right).$$

Now, for the other direction of the inequality, we will use an identical argument. First, tile a large square with (approximately) equal numbers of copies of $\mathcal{R}$ and $\mathcal{R}'$ (a vertically flipped copy of $\mathcal{R}$). Then, each tiling of the square yields a tiling of each triangle, so we have:

$$\sum_{\text{copies of } \mathcal{R}, \mathcal{R}'}\ln(\text{\# of tilings}) \leq (\text{total area}) \cdot S(\text{slope of } \Pi) + O\left(\varepsilon^2 \ln \tfrac{1}{\varepsilon}\right).$$

Noting that the lower bound applies for both the $\mathcal{R}$ and $\mathcal{R}'$ parts of the last sum, we obtain the desired bound. □

The final ingredient of the proof of the variational principle is contained in the following theorem:

Theorem 8.4   *Let $\mathcal{R}^*$ be a region with a piece-wise smooth boundary, and let h_b^* be a continuous height function on $\partial\mathcal{R}^*$. Further, let $\mathcal{R} = \mathcal{R}_L$ be a lattice region with a height function h_b on $\partial\mathcal{R}$, and for some fixed δ, suppose that*

[2]   We also need to sum over all possible choices of the approximately planar heights on the boundary of the square. However, the number of such choices is $\exp(O(L))$, and we can ignore this factor in the computation because we only care about the leading $O(L^2)$ asymptotics of the logarithm of the number of extensions.

[3]   Theorem 6.5 relies on Theorem 4.3, which was proven only for tori of specific sizes. However, the sizes form a positive-density subset, and thus it suffices to use only those sizes.

$\frac{1}{L}\mathcal{R}$ is within δ of $\mathcal{R}^*$ and $\frac{1}{L}h_b$ is within δ of h_b^*. Suppose also that there exist extensions of h_b to $\mathcal{R}$.

Then, given a function $h^* \in \mathcal{H}$ that agrees with h_b^* on $\partial\mathcal{R}^*$, we have:

$$\frac{1}{L^2}\ln(\textit{number of extensions of } h_b \textit{ to } \mathcal{R} \textit{ within } L\delta \textit{ of } Lh^*)$$

$$= \iint S(\nabla h^*)\, dx\, dy + o(1) \quad (8.3)$$

in the limit regime when we first send $L \to \infty$ and then $\delta \to 0$. The remainder $o(1)$ is uniform over the choice of h^*.

Proof Pick arbitrary (small) $\varepsilon > 0$. Pick $\ell = \ell(\varepsilon)$ as in Lemma 7.6 in Lecture 7 and create an ℓ-mesh of equilateral triangles on the region $\frac{1}{L}\mathcal{R}$. We will take $\delta < \ell\varepsilon$. Let $\tilde{h}$ be a height function that agrees with h^* on the vertices of the mesh and is piece-wise linear within each triangle of the mesh. By Lemma 7.6, on all but an ε fraction of the triangles, the function $\tilde{h}$ is within $\varepsilon\ell$ of h^* on the boundaries of the triangle.

Note that we can ignore the ε fraction of "bad" triangles because the contribution from these to (8.3) is negligible as $\varepsilon \to 0$. The negligibility is based on the observation that for each bad triangle, there is at least one discrete height function extending to the entire triangle the values of Lh^* on the vertices of this triangle in such a way that it stays within $L\delta$ of Lh^*. On the other hand, the total number of such extensions is not more than $\exp(O(\text{area of the triangle}))$.

Now, in order to lower bound the left-hand side of (8.3), note that we can obtain a valid extension of h_b to $\mathcal{R}$ by first fixing the values at the vertices of the mesh to agree with h^*, then choosing the values of the height function on the boundaries of the triangles in an arbitrary way (but keeping them close to the piece-wise linear approximation $\tilde{h}$) and, finally, counting all possible extensions from the boundary of the triangles of the mesh to the interior. Using Proposition 8.3, we get an estimate for the total number of tilings constructed in this way that matches the right-hand side of (8.3). Most of the height functions of these tilings are in $L\delta$-neighborhood of Lh^* because we chose them in such a way as the boundaries of the triangles and then can use the comparison with a torus, as in the proof of Proposition 8.1, to conclude that the height function is close to $\tilde{h}$ inside the triangles. Because $\tilde{h}$ is close to h^* by the construction, the desired lower bound is obtained.

For the upper bound of the RHS of (8.3), we take an arbitrary tiling of $\mathcal{R}$ with height function h and restrict it to each triangle in the mesh; note that this must yield a valid tiling of each good triangle. Thus, summing over all possible boundary conditions on the sides of the triangles of the mesh (whose normalized heights must be within $\ell\varepsilon$ of h^* because $\delta < \ell\varepsilon$) and using Proposition 8.3 for each triangle, we obtain the desired upper bound. It is crucial at this step that the number of choices of all possible boundary conditions on the sides of the triangles grows as $\exp(o(L^2))^4$ and is therefore negligible for (8.3).   □

Remark 8.5   One might want to be slightly more careful near the boundaries of the domain because h^* might not be defined in the needed points. We do not present these details here and refer the reader to Cohn et al. (2001). We also do not detail the uniformity of the remainder $o(1)$, leaving this as an exercise.

Proof of Theorem 5.15   For a function $h^* \in \mathcal{H}$ and $\delta > 0$, let $U_\delta(h^*)$ denote the set of all functions h from $\mathcal{H}$, for which $\sup_{(x,y)\in\mathcal{R}^*} |h(x,y) - h^*(x,y)| < \delta$.

Let $h^{\max}$ be the maximizer of Corollary 7.9. Choose an arbitrary small $\varepsilon > 0$. Using the uniqueness of the maximizer, the compactness of $\mathcal{H}$, and Theorem 7.5, we can find δ_0 (depending on ε), such that

$$\iint_{\mathcal{R}^*} S(\nabla h)\, dx\, dy < \iint_{\mathcal{R}^*} S(\nabla h^{\max})\, dx\, dy - \varepsilon,$$

for each $h \in \mathcal{H}$ outside $U_{\delta_0}(h^{\max})$.

Note that the choice $\delta_0 = \delta_0(\varepsilon)$ can be made in such a way that $\delta_0 \to 0$ as $\varepsilon \to 0$. Next, choose another $\delta > 0$ so that $o(1)$ in (8.3) is smaller in magnitude than $\varepsilon/2$. This δ can be arbitrarily small, and therefore we can require without loss of generality that $\delta < \delta_0$.

Using the compactness of the space $\mathcal{H}$, we can now choose finitely many elements $h_1^*, \ldots, h_k^* \in U_{\delta_0}(h^{\max})$ with $h_1^* = h^{\max}$ and finitely many elements $h_{k+1}^*, \ldots, h_m^* \in \mathcal{H} \setminus U_{\delta_0}(h^{\max})$ so that (for large L) each height function H_L of a tiling of $\mathcal{R}_L$ is in the δL neighborhood from one of the functions Lh_i^*, $1 \leq i \leq m$. The numbers k, m and functions $h_1^*, \ldots, h_m^*$ depend on ε and δ but not on L.

Let $\mathcal{N}_i$, $1 \leq i \leq m$, denote the number computed in (8.3) for $h^* = h_i^*$. Then we have a bound

$$\exp(L^2 \mathcal{N}_1) \leq \text{number of extensions of } h_b \text{ to } \mathcal{R}_L \leq \sum_{i=1}^{m} \exp(L^2 \mathcal{N}_i). \quad (8.4)$$

---

[4] There are $O(\ell^{-2})$ triangles in the mesh, and each of them has sides of length $O(\ell L)$. Hence, the total length of all sides of the triangles of the mesh is $O(\ell^{-1} \cdot L)$. If we define heights along the sides of triangles sequentially, starting from the boundary of the domain (where they are fixed), and then when we move to an adjacent vertex of the grid, we choose from not more than two possible values of the height function. Hence, the total number of possible values for the height function on all sides of all triangles can be upper-bounded by $2^{O(\ell^{-1}\cdot L)}$.

Sending $L \to \infty$ using (8.3) and the fact that $h_1^* = h^{\max}$ is the maximizer, we get

$$\iint_{\mathcal{R}^*} S(\nabla h^{\max}) \, dx \, dy - \varepsilon/2 \le \lim_{L \to \infty} \frac{1}{L^2} \ln (\text{number of extensions of } h_b \text{ to } \mathcal{R}_L)$$

$$\le \iint_{\mathcal{R}^*} S(\nabla h^{\max}) \, dx \, dy + \varepsilon/2. \tag{8.5}$$

Because $\varepsilon > 0$ is arbitrary, this proves (5.3).

Further, using $\delta < \delta_0$, we can write

$$1 - \mathbb{P}\left(H_L \text{ is within } 2L\delta_0 \text{ from } h^{\max}\right) \le \frac{\sum_{i=k+1}^{m} \exp(L^2 \mathcal{N}_i)}{\exp(L^2 \mathcal{N}_1)}. \tag{8.6}$$

Using (8.3) and the definition of δ_0, we conclude that the numerator in (8.6) is upper-bounded by

$$(m - k) \exp\left(L^2 \left(\iint_{\mathcal{R}^*} S(\nabla h^{\max}) \, dx \, dy - \varepsilon/2\right) + o(L^2)\right).$$

Simultaneously, using (8.3), the denominator is lower-bounded by

$$\exp\left(L^2 \iint_{\mathcal{R}^*} S(\nabla h^{\max}) \, dx \, dy + o(L^2)\right).$$

We conclude that the ratio in (8.6) goes to 0 as $L \to \infty$. Because δ_0 can be made arbitrarily small as $\varepsilon \to 0$, this implies convergence in probability $\frac{1}{L} H \to h^{\max}$. $\qquad \square$

9

Lecture 9: Euler–Lagrange and Burgers Equations

In the previous lectures we reduced the problem of finding the limit shape of random tilings to a variational problem. Now we start discussing ways to solve the latter.

9.1 Euler–Lagrange Equations

Recall that the limit shape of a tiling, under certain conditions, is given by:

$$\text{argmax}_h \iint S(\nabla h) \, dx \, dy.$$

We will consider the general problem of finding

$$\text{argmax}_h \iint F(h_x, h_y, h) \, dx \, dy \qquad (9.1)$$

for a general function F of three real arguments (and we are going to plug in h_x as the first argument, h_y as the second argument, and h as the third argument).

Proposition 9.1 (Euler–Lagrange Equations) *For maximal $h(x, y)$ solving (9.1), at all points (x, y) where h is smooth (as a function of (x, y)) and $F(h_x, h_y, h)$ is smooth (as a function of its argument in a neighborhood of $(h_x(x, y), h_y(x, y), h(x, y)))$, we have*

$$\frac{\partial}{\partial x} (F_1(h_x, h_y, h)) + \frac{\partial}{\partial y} (F_2(h_x, h_y, h)) - F_3(h_x, h_y, h) = 0, \qquad (9.2)$$

where F_i denotes the derivative of F in its ith argument.

In the noncoordinate form, the first two terms can be written as $\text{div}(\nabla F \circ \nabla h)$, representing the divergence of the vector field ∇F evaluated at $\nabla h(x, y)$.

Proof Note that because h is maximal, we must have, for all smooth g,

$$\frac{\partial}{\partial \varepsilon} \left(\iint F((h + \varepsilon g)_x, (h + \varepsilon g)_y, (h + \varepsilon g)) \, dx \, dy \right) \Bigg|_{\varepsilon=0} = 0.$$

Using the Taylor series expansions of F, this reduces to:

$$\iint \left(F_1(h_x, h_y, h)g_x + F_2(h_x, h_y, h)g_y + F_3(h_x, h_y, h)g \right) \, dx \, dy = 0.$$

Now pick g to be smooth, with compact support strictly inside the domain, and integrate by parts in x in the first term and in y in the second term to get the following:

$$\iint \left(\frac{\partial}{\partial x}(F_1(h_x, h_y, h)) + \frac{\partial}{\partial y}(F_2(h_x, h_y, h)) - (F_3(h_x, h_y, h)) \right) g \, dx dy = 0.$$

Because g is arbitrary, the first factor in the integrand has to vanish. $\square$

Specializing the Euler–Lagrange equations to our functional $S(\nabla h)$, we see two different situations:

- If ∇h at a point (x, y) lies strictly inside of the triangle of possible slopes, then (9.2) is a meaningful equation at this point.
- If ∇h is a boundary slope at (x, y), then (9.2) does not hold at such point because $S(\nabla h)$ fails to be a smooth function of ∇h; in fact, it is not even defined if we deform ∇h outside the triangle of possible slopes.

In the first situation, we say that (x, y) belongs to the *liquid region*, whereas the second one corresponds to the *frozen region*.

9.2 Complex Burgers Equation via a Change of Coordinates

If we try to write down the Euler–Lagrange equations directly for the functional of Theorem 5.15, we get a heavy and complicated expression. Kenyon and Okounkov (2007) found a way to simplify it by encoding the slope of the limit shape by a complex number. We now present their approach.

First, we wish to fix a coordinate system so that the gradient of h represents the probabilities of encountering each type of lozenge. Rather than using a rectangular coordinate system, we follow the conventions of Figure 3.2: the y-coordinate axis points up, and the x-coordinate axis points down and to the right. In other words, we use the up and right vectors from Figure 7.1 as two coordinate directions.

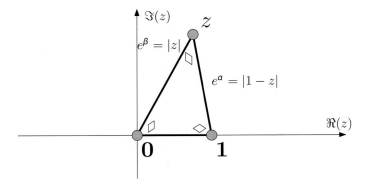

Figure 9.1 Triangle defining the complex slope z.

After fixing the x and y directions, we further modify the definition of the height function to

$$\tilde{h} = \frac{x + y - h}{3}. \tag{9.3}$$

Recall that by the definition of a height function (see discussion in Section 7.2.1), we have $h_x = p_\diamond + p_\searrow - 2p_\square$, and $h_y = p_\searrow + p_\square - 2p_\diamond$. Thus, because the local densities of the three types of lozenges sum to 1, we have the following for the modified height function:

$$\nabla \tilde{h} = (p_\square, p_\diamond). \tag{9.4}$$

Recall the results of Lecture 6, namely, Theorems 6.1 and 6.5 and the Legendre duality used to prove them. An important role is played by a triangle with side lengths e^α, e^β, 1. Its angles are $\pi(p_\diamond, p_\square, p_\searrow)$, with the correspondence shown in Figure 9.1. Let us position this triangle in the complex plane in such a way that two its vertices are 0 and 1, whereas the third one is a complex number z in the upper half-plane. Following Kenyon and Okounkov, we call z the "complex slope" encoding the triplet $(p_\diamond, p_\square, p_\searrow)$.

Using the identification of (9.4) and recalling that $\sigma = -S$, the Legendre duality of Lecture 6 can be now restated as

$$-\frac{\partial S}{\partial \tilde{h}_x} = \alpha = \ln|1 - z|, \qquad -\frac{\partial S}{\partial \tilde{h}_y} = \beta = \ln|z|. \tag{9.5}$$

It turns out that z satisfies a first-order differential equation.

Theorem 9.2 (Kenyon and Okounkov, 2007) *In the liquid region (i.e., where $p_\diamond, p_\square, p_\searrow > 0$), there exists a function $z(x, y)$ taking values in the complex upper half-plane such that*

$$(\tilde{h}_x, \tilde{h}_y) = \frac{1}{\pi} \left(\arg z, -\arg(1 - z) \right), \tag{9.6}$$

and

$$\frac{-z_x}{1 - z} + \frac{z_y}{z} = 0. \tag{9.7}$$

Proof   It can be shown using general results on variational problems and partial differential equations that $\tilde{h}$ is analytic in the liquid region; this is a particular case of *Hilbert's 19th problem*, and we refer the reader to the introduction of De Silva and Savin (2010) for references and discussion in the context of random surfaces.

Then $\tilde{h}$ is differentiable, and (9.6) is a restatement of (9.4) and the geometric definition of the complex slope z given previously.

By the Euler–Lagrange equations for the variational problem of Theorem 5.15, we obtain the following using (9.5):

$$0 = \frac{\partial}{\partial y}\left(\frac{\partial S}{\partial \tilde{h}_y}\right) + \frac{\partial}{\partial x}\left(\frac{\partial S}{\partial \tilde{h}_x}\right) = -\frac{\partial}{\partial y} \ln|z| - \frac{\partial}{\partial x} \ln|1 - z|.$$

Note that this is the real part of (9.7). For the imaginary part, we need to show

$$\frac{\partial}{\partial y} \arg z + \frac{\partial}{\partial x} \arg(1 - z) = 0,$$

which is the identity $\tilde{h}_{xy} = \tilde{h}_{yx}$ rewritten using (9.6).   □

We remark that the definition of the complex slope z is not canonical because we need to somehow identify three angles of the $(0, 1, z)$ triangle with three proportions $(p_\diamond, p_\square, p_\searrow)$, and there are six ways to do it. In particular, if we perform the substitution $u = \frac{z}{1-z}$ (this corresponds to a rearrangement of the vertices of the triangle of Figure 9.1), then the chain rule gives

$$u_x = \frac{z_x}{(1 - z)^2}, \quad u_y = \frac{z_y}{(1 - z)^2}.$$

In u-coordinates, (9.7) becomes the following:

$$u_x \cdot u = u_y. \tag{9.8}$$

This equation is well known and is called the "complex inviscid Burgers equation."

Exercise 9.3   There are six ways to identify three angles of the $(0, 1, z)$ triangle with three proportions $(p_\diamond, p_\square, p_\searrow)$. Two of these ways lead to equations (9.7) and (9.8). Find the equations corresponding to the four other ways.

9.3 Generalization to q^{Volume}-Weighted Tilings

We now consider a generalization of the uniformly random tilings by changing the weight of the measure to a q^{Volume}, where the volume is the number of unit cubes that must be added to the minimial tiling to obtain a given tiling.[1] The addition of one cube increases the height function $h(x, y)$ of Section 1.4 by 3 at a single point; hence, the volume is given by the double integral $\frac{1}{3} \iint h(x, y) \, dx \, dy$ up to a constant shift. In terms of the modified height function $\tilde{h}$ of (9.3), the addition of one cube decreases the height by 1, and therefore the volume is equal to *minus* the double integral: volume $= -\iint \tilde{h}(x, y) \, dx \, dy + \text{const}$.

Let us investigate the limit shape of the random height function under the measure q^{Volume} with q changing together with the linear size L of the domain by $q = e^{c/L}$.

Proposition 9.4 *Consider random tilings under the q^{Volume} measure with $q = \exp(c/L)$ as the linear size of the domain L grows. In the setting of Theorem 5.15, the limit shape for the height function still exists in this case; it is given by the following:*

$$h^* = \operatorname{argmax}_h \iint_{\mathcal{R}^*} \left(S(\nabla h) + \frac{c}{3} h \right) dx \, dy. \qquad (9.9)$$

Proof The proof is identical to the proof of the variational principle of Theorem 5.15 in the uniform case; we only need to multiply by q^{Volume} when we count tilings. □

Corollary 9.5 *The limit shape for q^{Volume} random lozenge tilings with $q = \exp(c/L)$ satisfies*

$$\operatorname{div}(\nabla S \circ \nabla h) = \frac{c}{3}, \qquad (9.10)$$

where $\circ$ is the evaluation operator; that is, ∇S is being evaluated at ∇h, and then div is applied to the resulting function of (x, y).

Proof This is the Euler–Lagrange equation for the variational problem (9.9). □

Theorem 9.2 extends to the q-weighted case with a similar proof.

Corollary 9.6 *For q^{Volume}-weighted tilings with $q = \exp(c/L)$, the statement of Theorem 9.2 remains valid with (9.7) replaced by*

$$\frac{-z_x}{1 - z} + \frac{z_y}{z} = c. \qquad (9.11)$$

---

[1] The q^{Volume} measure will appear again in Lectures 22–23 and then in Lecture 25. In particular, the effect of adding one cube on the height function is shown in Figure 25.2.

Proof The imaginary part of (9.11) is the same as that of (9.7), and its validity follows from $\tilde{h}_{xy} = \tilde{h}_{yx}$ relation. For the real part, we need to add an additional term in the Euler–Lagrange equations that produces the right-hand side of (9.11). Note that c is being multiplied by (-3) when we pass from h to $\tilde{h}$ by (9.3), hence the final form of (9.11). $\qquad\square$

9.4 Complex Characteristics Method

Because (9.11) is a first-order partial differential equation, it can be solved using the *method of characteristics*. The only tricky point is that the coordinate z is complex, whereas the textbooks usually present the method in the real case. Yet as we will see in this and the next lecture, all the principles continue to work in the complex case.

For simplicity, imagine first that z is real. Consider any curve $x(t), y(t), z(t)$ satisfying the following differential equations:

$$\frac{dx}{dt}(z-1) = \frac{dy}{dt}z = \frac{dz}{dt}\cdot\frac{1}{c}.$$

We call such a curve a "characteristic curve."

Proposition 9.7 *A surface $z = z(x, y)$ solving (9.11) must be a union of characteristic curves.*

Proof The normal vector to such a surface is $(z_x, z_y, -1)$. The tangent to a characteristic curve is, up to scaling by a function of t, $(\frac{1}{z-1}, \frac{1}{z}, c)$. Note that these two vectors have dot product 0 by (9.11), so the tangent vector to the characteristic curve is tangent to the surface. Thus, the characteristic curve remains on the surface. $\qquad\square$

Next, we solve the equation determining the characteristic curves. Choose $z = t$. Then we have

$$\frac{dy}{dz} = \frac{1}{cz}, \qquad \frac{dx}{dz} = \frac{1}{c(z-1)}.$$

The solutions to these equations are as follows:

$$cy = \ln z + k_1, \qquad cx = \ln(z-1) + k_2.$$

Rearranging, we get the following form for a characteristic curve:

$$ze^{-cy} = \tilde{k}_1, \qquad (1-z)e^{-cx} = \tilde{k}_2,$$

where $\tilde{k}_1$ and $\tilde{k}_2$ are arbitrary constants. The choice of these constants is related to the specification of the boundary conditions for the partial differential equation (9.11), and we need to specify which curves to use to form the surface. In general, a two-dimensional surface is a union of a one-parameter family of the curves; hence, we can parameterize the desired $\tilde{k}_1, \tilde{k}_2$ by the equation $Q(\tilde{k}_1, \tilde{k}_2) = 0$. We conclude that the solution to (9.11) must be given by $Q(ze^{-cy}, (1-z)e^{-cx}) = 0$ for some function Q.

We proceed to the complex case and seek a solution in the same form. The only difference is that Q is no longer arbitrary, but it is an *analytic function*. We have arrived at the following statement from Kenyon and Okounkov (2007):

Theorem 9.8 *Let $c \neq 0$, and take any analytic function of two variables $Q(u, v)$. The surface given by*

$$Q(ze^{-cy}, (1-z)e^{-cx}) = 0, \tag{9.12}$$

solves (9.11).

Proof Taking the derivatives in x and y of both sides of (9.12), we get

$$Q_1 z_x e^{-cy} + Q_2(-z_x)e^{-cx} - cQ_2(1-z)e^{-cx} = 0,$$

$$Q_1 z_y e^{-cy} - cQ_2 z e^{-cy} + Q_2(-z_y)e^{-cx} = 0,$$

where Q_i denotes the derivative of Q in its ith argument.

Note that we can isolate $\frac{Q_1 e^{-cy}}{Q_2 e^{-cx}}$ in both these expressions to obtain the following equality:

$$\frac{c(1-z) + z_x}{z_x} = \frac{z_y}{z_y - cz}.$$

Clearing denominators and rearranging, we precisely obtain (9.11). □

A remark is in order. The Burgers equation (9.11) is only valid in the liquid region. In the frozen part of the limit shape, only one type of lozenge is present, and the definition of the complex slope z is meaningless. On the other hand, the limit shape is still analytic – it is in fact linear. However, nonanalyticity appears on the arctic boundary – the separation curve between the liquid and frozen phases; the behavior of the limit shape on the two sides of this curve is very different.

We still have not discussed how to find Q based on the boundary conditions for lozenge tilings. We examine this question in the next lecture.

Lecture 10: Explicit Formulas for Limit Shapes

In the previous lectures we discussed two properties of the limit shapes of tilings: they solve a variational problem, and in liquid regions they solve the complex Burgers equation. In this lecture we concentrate on algorithmic ways to explicitly identify the limit shapes.

10.1 Analytic Solutions to the Burgers Equation

First, we continue our study of the solutions to the complex Burgers equation appearing as the Euler–Lagrange equation for the limit shapes of random tilings. The equation is as follows:

$$\frac{-z_x}{1-z} + \frac{z_y}{z} = c. \tag{10.1}$$

In the last lecture we have shown in Theorem 9.8 that an analytic function of two variables $Q(u, v)$ gives rise to a solution to (10.1). However, we have not explained whether every solution is obtained in this way, that is, whether we can use the formulation in terms of $Q(u, v)$ as an ansatz for finding all possible solutions to the Burgers equation and, hence, for finding the limit shapes of tilings. We proceed by showing that this is indeed the case (still in the situation $c \neq 0$).

Theorem 10.1 *If $c \neq 0$, given a solution to (10.1), there exists an analytic Q such that*

$$Q(ze^{-cy}, (1-z)e^{-cx}) = 0. \tag{10.2}$$

Remark 10.2 In the present formulation, Theorem 10.1 is a bit vague, and we are not going to make it more precise here. Of course, one can always choose Q to be identical zero, but this is not what we want. Rather, the theorem says that the first and second arguments of Q, ze^{-cy} and $(1 - z)e^{-cx}$, depend on each other (locally near each point (x_0, y_0) of the liquid region) in an analytic way.[1] The last property is easy to make precise in the case when the dependencies $(x, y) \rightarrow ze^{-cy}$ and $(x, y) \rightarrow (1 - z)e^{-cx}$ are nondegenerate; this is what we do in the proof.

For the applications, the statement of Theorem 10.1 should be treated as an ansatz: one should try to find a solution to (10.1) by searching for an analytic function Q. Often, Q will end up being very nice, like the polynomial in Theorem 10.7.

If we knew the exact shape of the liquid region, then once Q is found, it automatically provides the desired limit shape (maximizer) because of convexity of the functional in the variational problem of Theorem 5.15, which guarantees that a solution to the Euler–Lagrange equations is the global maximizer. However, the Euler–Lagrange equations or the complex Burgers equation do not tell us anything about the behavior (extremality) in the frozen regions, and therefore, additional checks in these regions are necessary after Q is identified; see Astala et al. (2020) and, in particular, Theorem 8.3 there for a more detailed discussion.

Sketch of the Proof of Theorem 10.1   Take $z(x, y)$ solving (10.1), and define $W = ze^{-cy}$ and $V = (1 - z)e^{-cx}$. What we need to prove is that $W(V)$ is an analytic function; in other words, $W(V)$ should satisfy the Cauchy–Riemann equations. This is equivalent to showing

$$\frac{\partial W}{\partial \bar{V}} = 0.$$

Here, we use the following notation: for a function $f(z) : \mathbb{C} \rightarrow \mathbb{C}$ with $z = x+iy$, we denote

$$\frac{\partial f}{\partial z} = \frac{1}{2}\left(\frac{\partial f}{\partial x} - i\frac{\partial f}{\partial y}\right),$$

and

$$\frac{\partial f}{\partial \bar{z}} = \frac{1}{2}\left(\frac{\partial f}{\partial x} + i\frac{\partial f}{\partial y}\right).$$

The following lemma is the first step in the computation:

Lemma 10.3   *The Jacobian of the map $(x, y) \rightarrow (V, W)$ vanishes; that is:*

$$W_x V_y = V_x W_y.$$

---

[1] Formally, this is a weaker property because in principle, local analytic dependence does not imply the existence of a global dependence as in (10.2); see the discussion in Astala et al. (2020), Remark 3.10.

Proof We have

$$W_x V_y = (z_x e^{-cy})(-z_y e^{-cx}),$$

and

$$V_x W_y = (-z_x e^{-cx} - c(1-z)e^{-cx})(z_y e^{-cy} - cze^{-cy}).$$

Therefore, subtracting and using (10.1), we find

$$V_x W_y - W_x V_y = e^{-cx} e^{-cy} (c^2 z(1-z) - cz_y(1-z) + cz_x z) = 0,$$

as desired. The cancellation may seem (almost) magical; however, the definitions of W and V were chosen by the method of characteristics to have this property. □

Now consider $W(V) = W(V(x, y)) = W(x, y)$. Using the chain rule, we find

$$\frac{\partial W}{\partial x} = \frac{\partial W}{\partial V}\left(\frac{\partial V}{\partial x}\right) + \frac{\partial W}{\partial \bar{V}}\left(\frac{\partial \bar{V}}{\partial x}\right),$$

and

$$\frac{\partial W}{\partial y} = \frac{\partial W}{\partial V}\left(\frac{\partial V}{\partial y}\right) + \frac{\partial W}{\partial \bar{V}}\left(\frac{\partial \bar{V}}{\partial y}\right).$$

Multiplying the first equation by $\frac{\partial V}{\partial y}$ and subtracting the second equation multiplied by $\frac{\partial V}{\partial x}$, we get a restatement of Lemma 10.3:

$$0 = \frac{\partial W}{\partial x}\frac{\partial V}{\partial y} - \frac{\partial W}{\partial y}\frac{\partial V}{\partial x} = \frac{\partial W}{\partial \bar{V}}\left(\frac{\partial \bar{V}}{\partial x}\frac{\partial V}{\partial y} - \frac{\partial \bar{V}}{\partial y}\frac{\partial V}{\partial x}\right).$$

This proves the desired Cauchy–Riemann relation $\frac{\partial W}{\partial \bar{V}} = 0$ unless we have that

$$\left(\frac{\partial \bar{V}}{\partial x}\frac{\partial V}{\partial y} - \frac{\partial \bar{V}}{\partial y}\frac{\partial V}{\partial x}\right) = 0. \tag{10.3}$$

The expression (10.3) coincides with the Jacobian of the transformation $(x, y) \to V$ from the (two-dimensional) liquid region to complex numbers (another two-dimensional real space), so as long as the mapping is not degenerate, we are done. Note that in a similar manner, we can demonstrate that $\frac{\partial V}{\partial W} = 0$, given a similar nondegeneracy for the map $(x, y) \to W$.

In other words, we have found $Q(\cdot, \cdot)$, which is analytic at any pair of complex points $(V(x, y), W(x, y))$ where at least one of the maps $(x, y) \to V(x, y)$, $(x, y) \to W(x, y)$ is nondegenerate. In a similar way, if we assume that $(x, y) \to \alpha V(x, y) + \beta W(x, y)$ is nondegenerate near (x_0, y_0) for some numbers α and β, then the possibility of choosing an analytic Q would also follow. □

Theorems 9.8 and 10.1 say that the map $(x, y) \mapsto (ze^{-cy}, (1 - z)e^{-cx})$ sends the liquid region to a one-dimensional complex Riemann surface in $\mathbb{C}^2$. In fact, this map is a bijection with its image, and therefore it can be used to equip the liquid region with the structure of a one-dimensional complex manifold.

Lemma 10.4 *In the liquid region, given* $(W, V) = (ze^{-cy}, (1 - z)e^{-cx})$, *one can recover* (x, y).

Proof Note that given W, V, one can immediately recover the arguments of the complex numbers z and $(1 - z)$ because x, y are real. Given this, one can recover z, and because $z \neq 0, 1$ because we are in the liquid region, we can therefore recover x, y. □

Next, we examine the $c = 0$ case.

Theorem 10.5 *For a solution* $z(x, y)$ *of the complex Burgers equation with* $c = 0$, *as in (9.7), there exists an analytic* Q_0 *such that*

$$Q_0(z) = yz + x(1 - z). \tag{10.4}$$

Remark 10.6 The discussion of Remark 10.2 also applies to Theorem 10.5. In addition, a solution $z = $ const is possible for (9.7) with specific boundary conditions. This corresponds to sending Q_0 to infinity in (10.4) so that the right-hand side becomes irrelevant.

The arguments leading to Theorem 10.5 are similar to the $c \neq 0$ case, and we omit all of this analysis except for the basic computation through the characteristics method.

The characterstic curves for

$$\frac{-z_x}{1 - z} + \frac{z_y}{z} = 0$$

solve

$$\frac{dx}{dt}(z - 1) = \frac{dy}{dt}(z),$$

and z is being kept constant. This gives $\frac{dx}{dy} = \frac{z}{z-1}$, leading to the parameterization of the curves by two arbitrary constant k_1, k_2 via

$$x = y\frac{z}{z - 1} + k_1, \qquad z = k_2.$$

This is equivalent to $x(1 - z) + yz = k_1(1 - k_2) = Q_0(z)$ in (10.4), where Q_0 expresses the dependence between k_1 and k_2.

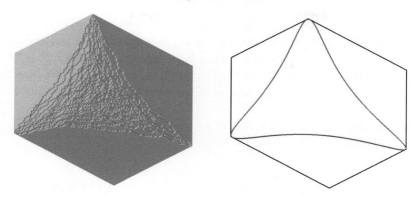

Figure 10.1 A q^{Volume}-random tiling of a $70 \times 90 \times 70$ hexagon with $q = 0.96$.

10.2 Algebraic Solutions

Theorems 10.1 and 10.5 claim the existence of an abstract Q that describes the liquid region. But how do we determine such Q? The following theorem of Kenyon and Okounkov shows that in the case of polygonal domains, Q is a polynomial:

Theorem 10.7 (Kenyon and Okounkov, 2007) *Take a polygonal region R with three-dimensional (3D) sides parallel to the coordinate direction repeated in cyclic order. (R is simply connected.) Then Q in Theorem 10.1 is a polynomial of degree (at most) d with real coefficients.*

We sketch a plan of the proof and leave it to the reader to figure out the necessary (highly nontrivial) details, following Kenyon and Okounkov (2007):

- Formulate the necessary properties that Q should have. "Q is given by a cloud curve." One of the properties is that the tangent vector should rotate d times as we move around the boundary of the frozen region.
- Study the $c = -\infty$ case. In this case, the limit shape is given by the tiling with the minimal height function.
- Obtain a solution (i.e., the function Q) for finite c through a continuous deformation from the $c = -\infty$ case, preserving the properties from the first step.
- Prove that the resulting Q is not only a solution to the Euler–Lagrange equations but a genuine maximizer. This check is tricky in the frozen regions because the Burgers equation no longer holds there.

Let us explain how the boundary of the frozen region can be computed using Q. In order to do this, we note that the boundary corresponds to real z such that $Q(ze^{-cy}, (1-z)e^{-cx}) = 0$. Because Q has real coefficients, z is a solution, and so is $\bar{z}$. Hence, the frozen boundary corresponds to the *double* real roots of Q. We conclude that the frozen boundary is the algebraic curve in e^{-cx}, e^{-cy} describing the values when $Q(ze^{-cy}, (1-z)e^{-cx})$ has a double root. We expect this curve to be tangent to all sides of the polygon that we tile; this is often enough to uniquely fix Q.

Let us now specialize to the case of a hexagon. Then, by Theorem 10.7, Q must be quadratic; that is, it has the form

$$Q(V, W) = a + bV + cW + dV^2 + eVW + fW^2,$$

and we need to substitute $W = ze^{-cy}$ and $V = (1-z)e^{-cx}$. This gives a quadratic equation in z. The boundary curve is then the curve where the discriminant of Q vanishes, which is a fourth-order polynomial equation in e^{-cx}, e^{-cy}. Because this curve needs to be tangent to the six sides of the hexagon,[2] this gives six equations in six unknowns a, b, c, d, e, f. Solving them, one finds Q.

Exercise 10.8   Find Q explicitly for the given proportions of the hexagon.

The results of a computer simulation of the q-weighted random tilings of the hexagon and corresponding theoretical boundary of the frozen boundary are shown in Figure 10.1. Random tilings were sampled using the algorithm from Borodin et al. (2010); another algorithm will be discussed in Lecture 25.

10.3 Limit Shapes via Quantized Free Probability

Let us now present another approach to the computation of the limit shapes of uniformly random lozenge tilings. Take any large domain, which does not have to be polygonal, but we assume that its boundary has one vertical segment with two adjacent diagonal segments forming a $2\pi/3$ angle with the vertical one, as in Figure 10.2.

The combinatorics imply that in the vicinity of this vertical segment, the horizontal lozenges $\diamond$ interlace, and at a vertical line at a distance N from this segment, we observe N lozenges $\diamond$. Let their (random) positions be $x_1, \ldots, x_N$ in the coordinate system with zero at the lower boundary of the domain.

---

[2] This tangency can be clearly seen in computer simulations; it is rigorously proven by a different method in Borodin et al. (2010). Also, the tangency condition is an essential component of the construction of Q in Kenyon and Okounkov (2007); see Theorem 3 there.

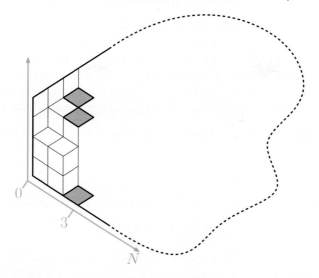

Figure 10.2 A part of a domain consisting of a vertical segment and two adjacent straight diagonal segments. At a distance N from the left boundary, we observe N horizontal lozenges.

We encode these $\diamond$ lozenges through their empirical measure μ_N and its Stieltjes transform $G_{\mu_N}(z)$:

$$\mu_N = \frac{1}{N} \sum_{i=1}^{N} \delta_{x_i/N}, \qquad G_{\mu_N}(z) = \int_{\mathbb{R}} \frac{1}{z-x} \mu_N(dx) = \frac{1}{N} \sum_{i=1}^{N} \frac{1}{z - x_i/N}.$$

Now assume that the domains start to grow, and set $N = \lfloor \alpha L \rfloor$. Here, L is the linear size of the domain, and parameter $\alpha > 0$ is chosen so that αL remains smaller than the length of the diagonal sides adjacent to the vertical segment; this choice guarantees that we still have N horizontal lozenges at a distance N from the left boundary.

The limit-shape theorem for uniformly random lozenge tilings (which we established through the variational principle in Lectures 5–8) implies the convergence

$$\lim_{L \to \infty} \mu_{\lfloor L\alpha \rfloor} = \mu_\alpha, \qquad \lim_{L \to \infty} G_{\mu_{\lfloor L\alpha \rfloor}}(z) = G_{\mu_\alpha}(z) = \int_{\mathbb{R}} \frac{1}{z-x} \mu_\alpha(dx),$$

where μ_α is a deterministic measure whose density encodes the asymptotic proportion of $\diamond$ lozenges along the vertical line at a distance α from the left vertical boundary of the rescaled domain. Note that μ_α can be directly related to the gradient of the limiting height function in the vertical direction.

The complex variable z in $G_{\mu_\alpha}(z)$ should be chosen outside the support of μ_α so that the integral is nonsingular.

Let us emphasize that x_i and the total weight of the measure are rescaled by N, rather than by L in the definition of μ_N – in other words, this rescaling depends on the horizontal coordinate α, which is slightly unusual.

It turns out that the measures μ_α are related to each other in a simple way.

Theorem 10.9 (Bufetov and Gorin, 2015)   *Define the* quantized R-transform *through*

$$R_{\mu_\alpha}^{quant}(z) = G_{\mu_\alpha}^{(-1)}(z) - \frac{1}{1 - e^{-z}}.$$

Then, for all $\alpha_1, \alpha_2 > 0$ for which the measures $\mu_{\alpha_1}, \mu_{\alpha_2}$ are defined, we have

$$\alpha_1 R_{\mu_{\alpha_1}}(z) = \alpha_2 R_{\mu_{\alpha_2}}(z). \tag{10.5}$$

Remark 10.10   By $G_{\mu_\alpha}^{(-1)}(z)$, we mean the functional inverse: The function $G_{\mu_\alpha}(z)$ behaves as $\frac{1}{z}$ as $z \to \infty$; hence, we can define an inverse function in a neighborhood of 0, which has a simple pole at 0. Subtraction of $\frac{1}{1-e^{-z}}$ cancels this pole, and $R_{\mu_\alpha}^{quant}(z)$ is an analytic function in a complex neighborhood of 0.

A modification of $R_\mu^{quant}(z)$ given by $R_\mu(z) := G_\mu^{(-1)}(z) - \frac{1}{z}$ is known in the free probability theory as the "Voiculescu R-transform"; see Voiculescu (1991), Voiculescu et al. (1992), and Nica and Speicher (2006). One standard use of the R-transform is to describe how the spectrum of large random Hermitian matrices changes when we add independent matrices or multiply them by projectors. Bufetov and Gorin (2015) demonstrated that $R_\mu^{quant}(z)$ replaces $R_\mu(z)$ when we deal with discrete analogues of these operations: compute tensor products and restrictions of the irreducible representations of unitary groups. The latter operation turns out to be intimately related to combinatorics of lozenge tilings (cf. Proposition 19.3 in Lecture 19), hence the appearance of $R_\mu^{quant}(z)$ in Theorem 10.9. We discuss the link between tilings and random matrices in more detail in Lectures 19 and 20.

From the computational point of view, Theorem 10.9 leads to the reconstruction of all μ_α (and hence of the limit shape) once such measure is known for some fixed choice of α. The following exercise provides an example where such an approach is useful.

Exercise 10.11   Consider the domain of Figure 10.3, which is a trapezoid with N teeth sticking out of the right boundary at every second position. By definition, as $N \to \infty$, the empirical measures on the right boundary converge to μ_1, which is a uniform measure on $[0, 2]$. Use this observation together with (10.5) to compute the limit shape as $N \to \infty$.

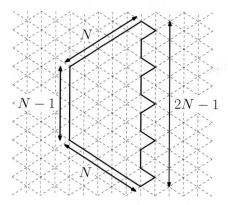

Figure 10.3 The half-hexagon domain: Trapezoid with N teeth sticking to the right at every second position.

In particular, Exercise 10.11 recovers the result of Nordenstam and Young (2011), which shows that the frozen boundary for this domain is a parabola.

The complex Burgers equation can also be extracted from the result of Theorem 10.9. The following exercise is close in spirit to the computations in Section 4 of Gorin (2017).

Exercise 10.12 Treating the formula (10.5) as a definition of the measures μ_α (assuming one of them to be given), show that the limit shape of tilings encoded by these measures satisfies the complex Burgers equation (9.7).

Hint: The density of the measure μ_α can be reconstructed as the imaginary part of the Stieltjes transform $G_{\mu_\alpha}(z)$ as z approaches the real axis. On the other hand, in the definition of the Kenyon–Okounkov complex slope, the density $p_\diamond$ becomes the argument of a certain complex number. This suggests the reconstruction of complex slope by exponentiating the Stieltjes transform.

We end this lecture by noting that in the random matrix theory to which we became linked through R-transforms, the complex Burgers equation has also shown up; see Matytsin (1994) and Guionnet (2004), which came before the work of Kenyon and Okounkov (2007).

11

Lecture 11: Global Gaussian Fluctuations for the Heights

11.1 Kenyon–Okounkov Conjecture

In previous lectures we found that the height function of random tilings for an arbitrary domain $\mathcal{R}$ converges in probability to a deterministic limit shape as the size of the domain grows to infinity:

$$\frac{1}{L}H(Lx, Ly) \xrightarrow{L \to \infty} h(x, y). \tag{11.1}$$

The limit shape $h(x, y)$ is a maximizer of a functional of the form

$$\iint_{\mathcal{R}} S(\nabla h) dx dy, \tag{11.2}$$

and it can also be identified with solutions to the complex Burgers equation on a complex function $z(x, y)$, where z encodes the gradient ∇h through a certain geometric procedure.

In this section we start discussing the asymptotic fluctuations of the random height function around its expectation. The fluctuations are very different in the liquid and frozen regions. In the latter, the height function is flat, with overwhelming probability for finite values of L, and hence, there are essentially no fluctuations. For the former, the situation is much more interesting.

Conjecture 11.1 (Kenyon and Okounkov, 2007; Kenyon, 2008) *Consider* q^{Volume}-*weighted random tilings in domain* $\mathcal{R}$ *of linear size* L *and with* $q = \exp(c/L)$. *In the liquid region* $\mathcal{L}$,

$$H(Lx, Ly) - \mathbb{E}[H(Lx, Ly)] \xrightarrow[L \to \infty]{d} \text{GFF}, \tag{11.3}$$

where GFF *is the Gaussian free field on* $\mathcal{L}$ *with respect to the complex structure given by* ze^{-cy} *and with Dirichlet boundary conditions. The* GFF *is normalized so that*

$$\text{Cov}\left(H(Lx, Ly), H(Lx', Ly')\right) \approx -\frac{1}{2\pi^2} \ln d\left((x, y), (x', y')\right) \qquad (11.4)$$

for $(x, y) \approx (x', y')$, *with* $d(\cdot, \cdot)$ *denoting the distance in the local coordinates given by the complex structure and with normalization for the definition of the height function as in the discrete version of* (9.3), (9.4).

In this lecture we aim to define and understand the limiting object – the Gaussian free field (GFF). The next lecture contains a heuristic argument toward the validity of Conjecture 11.1. Later on, in Lectures 22 and 23, we present a proof of a particular case of this conjecture for random plane partitions.

For general domains $\mathcal{R}$, Conjecture 11.1 is a major open problem in the area. Yet, there are several approaches to its proofs for particular classes of domains:

- One approach was suggested in Kenyon (2000, 2001, 2008). The correlation functions of tilings are minors of the inverse Kasteleyn matrix, which can be thought of as a discrete harmonic function. On the other hand, the covariance of GFF is a continuous harmonic function. Thus, one can be treated as a discretization of the other, which allows us to use various convergence results from the analysis. Technically, two challenging points of such an approach are dealing with nontrivial complex structures and dealing with frozen boundaries. Despite some progress (cf. Russkikh, 2018, 2019; Berestycki et al., 2020), the proofs along these lines until very recently were restricted to a special class of domains, which, in particular, have no frozen regions. At the time of writing of this book, new developments appeared in Chelkak et al. (2020) and Chelkak and Ramassamy (2020): the central idea is to construct a discrete approximation of the limiting complex structure using *origami maps*. There is a hope to reach a resolution of Conjecture 11.1 through this approach in the future.
- The second approach (see Borodin and Ferrari, 2014; Petrov, 2015) is to use a double contour integral representation for the inverse Kasteleyn matrix and then apply the steepest descent method for the asymptotic analysis. We will discuss some components of this method in Lectures 14–16. This method is naturally restricted to the domains where such contour integral representations are available. Yet, such domains include a class of specific polygons with many arbitrary sides.
- The third approach combines a gadget called "discrete loop equations" with the idea of applying differential/difference operators to inhomogeneous partition functions to extract asymptotic information. We discuss discrete

loop equations (cf. Borodin et al., 2017) in Lecture 21 and the use of difference operators for GFF asymptotics (cf. Borodin and Gorin, 2015; Bufetov and Gorin, 2018; Ahn, 2018) in Lectures 22–23. Tilings of non–simply connected domains with holes were analyzed through this approach in Bufetov and Gorin (2019).

- In some cases, the marginal distributions of random tilings can be identified with discrete log-gases, as we will compute in Lecture 19. Yet another approach to global fluctuations exists in this case and links it to the asymptotic of recurrence coefficients of orthogonal polynomials; see Breuer and Duits (2017) and Duits (2018).

We remark that our toy example of the uniform measure on tilings of the hexagon is covered by each of the last three methods.

11.2 Gaussian Free Field

We start our discussion of Conjecture 11.1 by defining the (two-dimensional) GFF in a domain $\mathcal{D} \subset \mathbb{C}$ in the *standard* complex structure of $\mathbb{C}$.

Take the Laplace operator on $\mathcal{D}$:

$$\triangle = \frac{\partial^2}{\partial x^2} + \frac{\partial^2}{\partial y^2}\bigg|_{(x,y)\in\mathcal{D}}.$$

We consider inverse operator $-\triangle^{-1}$ (note the minus sign) taken with Dirichlet (i.e., identical zero) boundary conditions on the boundary $\partial\mathcal{D}$. $\triangle^{-1}$ is an integral operator (with respect to the Lebesgue measure on $\mathcal{D}$), and its kernel is known as the "Green function." Equivalently, we define the following:

Definition 11.2 The Green function $G_{\mathcal{D}}(z, z')$ on $\mathcal{D}$ with Dirichlet boundary condition is a function $G \colon \mathcal{D} \times \mathcal{D} \to \mathbb{R}$ of two complex arguments $z = x + \mathbf{i}y$, $z' = x' + \mathbf{i}y'$, such that

$$\triangle G_{\mathcal{D}}(z, z') = -\delta(z = z'),$$

where $\triangle$ can be applied either in z or in z' variables, and

$$G_{\mathcal{D}}(z, z') = 0, \text{ if either } z \in \partial\mathcal{D} \text{ or } z' \in \partial\mathcal{D}.$$

We remark without proof that (for nonpathological $\mathcal{D}$) the Green function exists and is unique and symmetric (i.e., $G_{\mathcal{D}}(z, z') = G_{\mathcal{D}}(z', z)$). When $\mathcal{D}$ is unbounded, one should add a (at most) logarithmic growth condition at ∞ to the definition of the Green functions.

Exercise 11.3 Suppose that $\mathcal{D}$ is the upper half-plane $\mathbb{H} = \{z \in \mathbb{C} \mid \Im z > 0\}$. Show that in this case,

$$G_{\mathbb{H}}(z, z') = -\frac{1}{2\pi} \ln \left| \frac{z - z'}{z - \bar{z}'} \right|. \tag{11.5}$$

For general domains, there is often no explicit formula for the Green function, but locally, such a function always grows like a logarithm of the distance:

$$G_{\mathcal{D}}(z, z') \sim -\frac{1}{2\pi} \ln |z - z'| \quad \text{for } z \approx z'. \tag{11.6}$$

We can now present an *informal* definition of the GFF.

Definition 11.4 The Gaussian Free Field on $\mathcal{D}$ is a random Gaussian function $\mathrm{GFF}_{\mathcal{D}} : \mathcal{D} \to \mathbb{R}$ such that

$$\mathbb{E}[\mathrm{GFF}_{\mathcal{D}}] = 0,$$

$$\mathbb{E}[\mathrm{GFF}_{\mathcal{D}}(z)\mathrm{GFF}_{\mathcal{D}}(z')] = G_{\mathcal{D}}(z, z').$$

However, there is a problem with this definition: if $z = z'$, then $G_{\mathcal{D}}(z, z') = +\infty$; hence, the values of $\mathrm{GFF}_{\mathcal{D}}$ have infinite variance. This means that $\mathrm{GFF}_{\mathcal{D}}$ is not a function but a *generalized function*: we cannot ask about the values of this random function at a point, but we can compute its integrals (or pairings) with reasonably smooth test functions.

Definition 11.5 For a test function $u : \mathcal{D} \to \mathbb{R}$, such that

$$\iint_{\mathcal{D}} u(z)G(z, z')u(z') \, dx \, dy \, dx' \, dy' < \infty, \qquad z = x + \mathbf{i}y, \quad z' = x' + \mathbf{i}y',$$

we define a pairing of $\mathrm{GFF}_{\mathcal{D}}$ with u, as a mean zero Gaussian random variable $\langle \mathrm{GFF}_{\mathcal{D}}, u \rangle$ with variance

$$\mathbb{E}\left[\langle \mathrm{GFF}_{\mathcal{D}}, u \rangle \langle \mathrm{GFF}_{\mathcal{D}}, u \rangle \right] = \iint_{\mathcal{D} \times \mathcal{D}} u(z)G(z, z')u(z') \, dx \, dy \, dx' \, dy'.$$

If we take several such u's, then the pairings are jointly Gaussian with covariance

$$\mathbb{E}\left[\langle \mathrm{GFF}_{\mathcal{D}}, u \rangle \langle \mathrm{GFF}_{\mathcal{D}}, v \rangle \right] = \iint_{\mathcal{D} \times \mathcal{D}} u(z)G(z, z')v(z') \, dx \, dy \, dx' \, dy'. \tag{11.7}$$

Informally, one should think about the pairings as integrals:

$$\langle \mathrm{GFF}_{\mathcal{D}}, u \rangle = \iint_{\mathcal{D}} \mathrm{GFF}_{\mathcal{D}}(z, \cdot)u(z) \, dx \, dy. \tag{11.8}$$

From this point of view, one can take as u a (reasonably smooth) measure rather than a function.

By invoking the Kolmogorov consistency theorem, $\mathrm{GFF}_{\mathcal{D}}$ is well defined as a stochastic process through Definition 11.5 if we show that the joint distributions appearing in this definition are consistent. Because they are all mean 0 Gaussians, it suffices to check that the covariance is positive-definite.

Lemma 11.6   *The covariance of* $\langle \mathrm{GFF}_{\mathcal{D}}, u \rangle$ *given by* (11.7) *is positive-definite.*

Proof   We need to check that the operator with kernel $G_{\mathcal{D}}$ is positive-definite, which is equivalent to $-\triangle$ being positive-definite. The latter can be checked by integrating by parts to reduce to the norm of the gradient:

$$\langle -\triangle u, u \rangle = \iint_{\mathcal{D}} (-\triangle u(z)) u(z) \, dx \, dy = \iint_{\mathcal{D}} \|\nabla u(z)\|_{L^2}^2 \, dx \, dy > 0. \quad \square$$

Remark 11.7   A better alternative to Definition 11.5 would be to define a functional space (like the space of continuous functions in the definition of the Brownian motion) to which the $\mathrm{GFF}_{\mathcal{D}}$ belongs. We refer the reader to Sheffield (2007), Section 4 of Dubedat (2009), Section 2 of Hu et al. (2010), and Werner and Powell (2020) for more details on the definition of the GFF.

An important property of the GFF is its *conformal invariance*.

Recall that a map $\phi \colon \mathcal{D} \to \mathcal{D}'$ is called *conformal* if it is holomorphic, it is bijective, and its inverse is also holomorphic. A direct computation with Laplacian $\triangle$ leads to the conclusion that the Green function is preserved under conformal bijections, thus implying the following result:

Exercise 11.8   If ϕ is a conformal bijection of $\mathcal{D}$ with $\mathcal{D}'$, then

$$\{\mathrm{GFF}_{\mathcal{D}'}(\phi(z))\}_{z \in \mathcal{D}'} \overset{d}{=} \{\mathrm{GFF}_{\mathcal{D}}(z)\}_{z \in \mathcal{D}}.$$

For instance, in the case when the domain is an upper half-plane, $\mathcal{D} = \mathcal{D}' = \mathbb{H}$, the conformal bijections are Moebius transformations:

$$z \mapsto \frac{az+b}{cz+d}, \qquad a, b, c, d \in \mathbb{R},$$

and one readily checks that the formula in (11.5) is unchanged under such transformations.

Another remark is that if we follow the same recipe for defining the GFF in dimension 1, rather than 2, then we get the Brownian bridge.

Lemma 11.9   *Consider the standard Brownian bridge,*

$$B \colon [0, 1] \to \mathbb{R} \ , \qquad B(0) = B(1) = 0, \tag{11.9}$$

which can be defined as the Brownian motion conditioned to be at 0 *at time* 1, *or as a centered Gaussian process with covariance function*

$$\mathbb{E}B_s B_t = C(s,t) = \min(t,s)(1 - \max(t,s)), \qquad 0 \le t, s \le 1. \quad (11.10)$$

Then the function $C(s,t)$ is the kernel of the (minus) inverse Laplace operator on the interval $[0,1]$ with Dirichlet boundary conditions at 0 and 1.

Proof Because the Laplace operator in dimension 1 is simply the second derivative, and the vanishing of $C(s,t)$ at $t = 0$, $s = 0$, $t = 1$, and $s = 1$ is clear, it suffices to check that

$$\frac{\partial^2}{\partial t^2} C(s,t) = -\delta(s = t). \quad (11.11)$$

We compute

$$\frac{\partial C(s,t)}{\partial t} = \begin{cases} 1 - s, & t < s, \\ -s, & t \ge s. \end{cases}$$

Differentiating, we get (11.11). $\qquad\square$

We now give another *informal* definition of the GFF. According to it, GFF$_\mathcal{D}$ is a probability measure on functions f in $\mathcal{D}$ with 0 boundary conditions with density

$$\rho(f) \sim \exp\left(-\frac{1}{2}\int_\mathcal{D} \|\nabla f\|_{L^2}^2 \, dx \, dy\right). \quad (11.12)$$

The formula in (11.12) is based on the well-known computation for the finite-dimensional Gaussian vectors.

Exercise 11.10 For a Gaussian vector $v \in \mathbb{R}^N$ with probability density

$$\rho(v) = \frac{\sqrt{\det B}}{\sqrt{(2\pi)^N}} \exp\left(-\frac{1}{2}\langle v, Bv\rangle\right),$$

the covariance matrix $\mathbb{E}[vv^T]$ is equal to B^{-1}.

Because the covariance for GFF$_\mathcal{D}$ is given by the (minus) inverse Laplace operator, the matrix B appearing in its density should be identified with $-\triangle$. Integrating by parts, we then arrive at the formula in (11.12). We will repeatedly use (11.12) in the heuristic arguments of the following lectures. However, let us emphasize that the precise mathematical meaning of (11.12) is very tricky: it is unclear with respect to which underlying measure the density is computed, and also the symbol $\|\nabla f\|_{L^2}^2$ is not well defined for nonsmooth f (and we already know that GFF is not smooth; even its values at points are not properly defined).

11.3 Gaussian Free Field in Complex Structures

The GFF appearing in Conjecture 11.1 is defined using the complex structure ze^{-cy} rather than the standard complex structure of the plane. In this section, we explain the meaning of this statement in two ways.

For the first approach, recall the results of Theorems 9.8 and 10.1 and Lemma 10.4. There exists an analytic function $Q(v, w)$, such that the map

$$(x, y) \mapsto (ze^{-cy}, (1 - z)e^{-cx}) \tag{11.13}$$

is a bijection of the liquid region $\mathcal{L}$ with (a part) of the curve $Q(v, w) = 0$ in $\mathbb{C}^2$. This part, being embedded in $\mathbb{C}^2$, has a natural local coordinate system on it; in particular, we can define the gradient and the Laplace operator on the curve $Q(v, w) = 0$ and then use one of the definitions of GFF from the previous section directly on this curve. Then, we pull back the resulting field onto the liquid region $\mathcal{L}$ using the map (11.13). This is the desired field in Conjecture 11.1.

For the second approach, given the liquid region $\mathcal{L}$ and the complex slope $z = z(x, y)$ on it, we define a class of analytic functions. We give two closely related definitions:

Definition 11.11 $f : \mathcal{L} \to \mathbb{C}$ is analytic with respect to ze^{-cy} if for each $(x_0, y_0) \in \mathcal{L}$, there exists a holomorphic function g of a complex variable, such that

$$f(x, y) = g(ze^{-cy}) \tag{11.14}$$

in a small neighborhood of (x_0, y_0).

Definition 11.12 $f : \mathcal{L} \to \mathbb{C}$ is analytic with respect to ze^{-cy} if f satisfies the following first-order partial differential equation (PDE):

$$\frac{f_y}{z} - \frac{f_x}{1 - z} = 0. \tag{11.15}$$

The complex Burgers equation of Corollary 9.6 implies that ze^{-cy} itself satisfies (11.15):

$$\frac{(ze^{-cy})_y}{z} - \frac{(ze^{-cy})_x}{1 - z} = 0. \tag{11.16}$$

This can be used to show that (11.14) implies (11.15). If ze^{-cy} is locally injective, then we can also argue in the opposite direction: (11.15) implies the Cauchy–Riemann relations for the function g in (11.14), and the two definitions become equivalent.

All of these approaches accomplish the same goal: they create a structure of a one-dimensional complex manifold (Riemann surface) on the liquid region $\mathcal{L}$.

The general *Riemann uniformization theorem* says that all Riemann surfaces of the same topology are conformally equivalent. In our case, it specifies to the following:

Theorem 11.13 *There exists a conformal bijection* $\Omega\colon \mathcal{L} \mapsto \mathcal{D}$ *with some domain* $\mathcal{D} \subset \mathbb{C}$, *where* $\mathcal{D}$ *is equipped with the standard complex structure of* $\mathbb{C}$, *and the liquid region* $\mathcal{L}$ *is equipped with the complex structure of* ze^{-cy}; *that is,* Ω *is analytic by Definitions 11.11 and 11.12.*

In other words, Ω is an identification of (part of) the curve $\{Q(v, w) = 0\} \subset \mathbb{C}^2$ with a part of the plane $\mathbb{C}$.

Definition 11.14 The GFF in the liquid region $\mathcal{L}$ is defined through

$$\mathrm{GFF}_{\mathcal{L}}(x, y) = \mathrm{GFF}_{\mathcal{D}}(\Omega(x, y)).$$

This means that for test functions u on $\mathcal{L}$, its pairings with the Gaussian Free Field $\langle \mathrm{GFF}_{\mathcal{L}}, u \rangle$ are jointly Gaussian with covariance

$$\mathbb{E}\left[\langle \mathrm{GFF}_{\mathcal{L}}, u \rangle \langle \mathrm{GFF}_{\mathcal{L}}, v \rangle\right] = \mathbb{E}\left[\langle \mathrm{GFF}_{\mathcal{D}}, [u \circ \Omega^{-1}] \cdot \mathbf{J}_{\Omega^{-1}} \rangle \langle \mathrm{GFF}_{\mathcal{D}}, [u \circ \Omega^{-1}] \cdot \mathbf{J}_{\Omega^{-1}} \rangle\right]$$

$$= \int\!\!\int_{\mathcal{D} \times \mathcal{D}} [u \circ \Omega^{-1}](z) G_{\mathcal{D}}(z, z') [v \circ \Omega^{-1}](z')\, \mathbf{J}_{\Omega^{-1}}(z) \mathbf{J}_{\Omega^{-1}}(z')\, dx\, dy\, dx'\, dy',$$

$$(11.17)$$

where $\mathbf{J}_{\Omega^{-1}}$ is the Jacobian of the map Ω^{-1}, and $z = x + \mathbf{i}y$, $z' = x' + \mathbf{i}y'$.

Let us emphasize that under the map Ω, GFF is transformed as a function (rather than a distribution or measure); this again indicates that the test functions u should be treated as measures $u(x + \mathbf{i}y)\, dx\, dy$.

One might be worried that the map Ω in Theorem 11.13 is not unique because it can be composed with any conformal bijection $\phi\colon \mathcal{D} \mapsto \mathcal{D}'$. However, the conformal invariance of the GFF of Exercise 11.8 guarantees that the distribution of $\mathrm{GFF}_{\mathcal{L}}$ does not depend on the choice of Ω.

Finally, note that (11.4) differs from (11.6) by a factor of π. This is not a typo but, rather, a universal constant appearing in tilings. We will see this constant appearing in the computation in the next section – the conceptual reasons for the particular value of the constant for random lozenge tilings are not completely clear to us at this point. The same factor π appears in Theorem 4.5 of Kenyon et al. (2006) (see also Section 5.3.2) as a universal normalization prefactor for random tilings with *periodic weights*.

12

Lecture 12: Heuristics for the Kenyon–Okounkov Conjecture

This lecture focuses on a *heuristic* argument for the Kenyon–Okounkov conjecture on the convergence of the centered height function to the Gaussian free field (GFF). We concentrate on only the case $c = 0$ (i.e., the uniform measure) here. By the variational principle (Cohn et al., 2001), in the limit, most height functions will be close to the unique maximizer h^*, which we have pinned down (somewhat) explicitly in Lecture 9 by showing that the normalized shifted height function $\tilde{h}^*$ satisfies $(\tilde{h}^*_x, \tilde{h}^*_y) = \frac{1}{\pi}(\arg(z), -\arg(1-z))$, where $z = z(x, y)$ satisfies the transformed Burgers equation:

$$-\frac{z_x}{1-z} + \frac{z_y}{z} = 0. \tag{12.1}$$

The Kenyon–Okounkov conjecture focuses on the fluctuations about this limit shape, and we repeat it now in the $c = 0$ case.

Conjecture 12.1 (Kenyon and Okounkov, 2007; Kenyon, 2008) *Take the same setup as in Theorem 5.15: let R^* be a domain in $\mathbb{R}^2$ with the piece-wise smooth boundary ∂R^* and a specified boundary height function h_b. Take a sequence of domains R_L for a sequence $L \to \infty$, such that $\frac{\partial R_L}{L} \to \partial R^*$ and the boundary height functions converge as well. Now, for each L, consider the uniform probability measure on tilings of R_L. Finally, let $H(Lx, Ly)$ be the value of the height function of a random tiling of R_L at (Lx, Ly). Then, for the points $(x, y) \in R^*$ in the liquid region, as $L \to \infty$, the centered heights $H(Lx, Ly) - \mathbb{E}[H(Lx, Ly)]$ converge in distribution to the GFF in the complex structure given by z.*[1]

Remark 12.2 Let us clarify the normalizations. If we choose the (modified) definition of the height function as in (9.3) and (9.4), then $\sqrt{\pi}(H(Lx, Ly) - \mathbb{E}[H(Lx, Ly)])$ should asymptotically behave like the GFF of Definition 11.14 with the short-scale behavior of the covariance as in (11.6).

---

[1] See the previous lecture for a precise description of what this means.

From here on, we will denote the liquid region by $\mathcal{L}$. Recall from previous lectures that for large L, for a height function H on R_L,

$$\Pr(H) \approx \exp\left(L^2 \iint_{\mathcal{L}} S(\nabla H)dxdy\right). \tag{12.2}$$

(Note that $\Pr(H)$ is really describing not a probability but a probability density at H in the space of height functions, i.e., the probability of being close to H.)

Because we would like to study limiting fluctuations about h^*, we can let $H = h^* + L^{-1}g$, where g is a random function supported in the liquid region that represents the fluctuations about the limiting height function h^*. Plugging this H into (12.2) yields

$$\Pr(h^* + L^{-1}g) \approx \exp\left(L^2 \iint_{\mathcal{L}} S(\nabla h^* + L^{-1}\nabla g)dxdy\right). \tag{12.3}$$

We now Taylor expand the integrand $f(L^{-1}) := S(\nabla h^* + L^{-1}\nabla g)$ with respect to the variable L^{-1},

$$f(L^{-1}) = f(0) + f'(0)L^{-1} + \frac{1}{2}f''(0)L^{-2} + \text{(lower-order terms)}.$$

Using the chain rule, we have

$$f(0) = S(\nabla h^*),$$

$$f'(0) = S_1(\nabla h^*)g_x + S_2(\nabla h^*)g_y,$$

where S_1, S_2 are the partial derivatives of S with respect to its first and second coordinate, viewing S as a function of two arguments h_x^*, h_y^* – the two coordinates of ∇h^*. We also have

$$f''(0) = S_{11}(\nabla h^*)(g_x)^2 + 2S_{12}(\nabla h^*)g_x g_y + S_{22}(\nabla h^*)(g_y)^2.$$

The first term yields an L-dependent normalizing constant $\exp\left(L^2 \iint_{\mathcal{L}} S(\nabla h^*)\right)$, which does not affect the probabilities. Because $H = h^*$ minimizes

$$\iint_{\mathcal{R}^*} S(\nabla H) \, dxdy,$$

it follows that

$$\left.\frac{d}{dt}\right|_{t=0} \iint S(\nabla(h^* + tg)) \, dxdy = 0; \tag{12.4}$$

this is exactly what was used to derive the Euler–Lagrange equations. Hence, the integral of the $f'(0)L^{-1}$ term vanishes. The L^{-2} in the quadratic term cancels with the L^2 outside the integral in (12.3), yielding a nonzero probability density function at g.

Thus,

$$\Pr(h^* + L^{-1}g) \approx \exp\left(L^2 \iint_{\mathcal{L}} S(\nabla h^* + L^{-1}\nabla g)\, dx dy\right)$$
$$= \exp\left(\frac{1}{2}\iint_{\mathcal{L}} \left[S_{11}(\nabla h^*)(g_x)^2 + 2S_{12}(\nabla h^*)g_x g_y + S_{22}(\nabla h^*)(g_y)^2\right] dx dy \right.$$
$$\left. + \text{(lower order)}\right). \quad (12.5)$$

Having completed this heuristic derivation, we can now (rigorously) show that the right-hand side of (12.5) is exactly what we would get if the height-function fluctuations converged to a GFF. This is the content of the following lemma:

Lemma 12.3   *With the previous setup, we have*

$$\iint_{\mathcal{L}} \left[S_{11}(\nabla h^*)(g_x)^2 + 2S_{12}(\nabla h^*)g_x g_y + S_{22}(\nabla h^*)(g_y)^2\right] dx dy$$
$$= 2\pi \mathbf{i} \iint_{\mathcal{L}} \frac{dg}{dz}\frac{dg}{d\bar{z}}\, dz \wedge d\bar{z}. \quad (12.6)$$

Before proving this lemma, we show why, when combined with (12.5), it shows that the fluctuations converge to a GFF. Recall from (11.12) in Lecture 11 that the GFF on a domain D may be thought of as a probability measure on functions on D that vanish on ∂D, with density

$$\rho(f) \sim \exp\left(-\frac{1}{2}\iint_D ||\nabla f||^2 dx dy\right). \quad (12.7)$$

Because $dz = dx + \mathbf{i}dy$ and $d\bar{z} = dx - \mathbf{i}dy$,

$$dz \wedge d\bar{z} = -2\mathbf{i}dx \wedge dy. \quad (12.8)$$

Similarly,

$$\frac{dg}{dz} = \frac{1}{2}\left(\frac{\partial g}{\partial x} - \mathbf{i}\frac{\partial g}{\partial y}\right), \quad \text{and} \quad \frac{dg}{d\bar{z}} = \frac{1}{2}\left(\frac{\partial g}{\partial x} + \mathbf{i}\frac{\partial g}{\partial y}\right), \quad (12.9)$$

so that

$$\frac{dg}{dz}\frac{dg}{d\bar{z}} = \frac{1}{4}||\nabla g||^2. \quad (12.10)$$

Making these substitutions in Lemma 12.3 and combining with (12.5), we have that the probability density on height function fluctuations g converges to the one in (12.7). This completes the heuristic argument for GFF fluctuations; the remainder of this lecture will be spent proving Lemma 12.3.

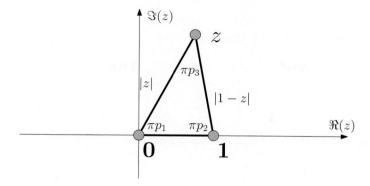

Figure 12.1 Correspondence between complex slope z and local proportions of lozenges identified with angles of a triangle.

Proof of Lemma 12.3   First, recall from Lecture 9 that

$$S_1(p_1, p_2, p_3) = \ln|1 - z|, \qquad (12.11)$$

$$S_2(p_1, p_2, p_3) = \ln|z|, \qquad (12.12)$$

where z is the complex number encoding the probabilities p_1, p_2, and $p_3 = 1 - p_1 - p_2$, which we use as shorthand for the three lozenge probabilities $p_⬨(\cdot)$, $p_⬦(\cdot)$, and $p_◇(\cdot)$.

From Figure 12.1, clearly $z = |z|e^{i\pi p_1}$, and by the law of sines

$$\frac{|z|}{\sin \pi p_2} = \frac{1}{\sin \pi(1 - p_1 + p_2)} = \frac{1}{\sin \pi(p_1 + p_2)}.$$

Therefore,

$$z = \frac{\sin \pi p_2}{\sin \pi(p_1 + p_2)} e^{i\pi p_1}.$$

It follows that

$$S_{12} = \frac{\partial \ln|z|}{\partial p_1} = \frac{\partial}{\partial p_1}(\ln(\sin \pi p_2) - \ln(\sin \pi(p_1 + p_2))) = -\pi \cot \pi(p_1 + p_2),$$

$$S_{22} = \frac{\partial \ln|z|}{\partial p_2} = \pi \cot \pi p_2 - \pi \cot \pi(p_1 + p_2),$$

$$S_{11} = \pi \cot \pi p_1 - \pi \cot \pi(p_1 + p_2),$$

where the formula for S_{11} follows by symmetry of p_1 and p_2 from the formula for S_{22}.[2] Recall that g is a function of x, y and also of $(z, \bar{z})$; hence,

$$\frac{\partial g}{\partial x} = \frac{dg}{dz} z_x + \frac{dg}{d\bar{z}} \bar{z}_x, \tag{12.13}$$

$$\frac{\partial g}{\partial y} = \frac{dg}{dz} z_y + \frac{dg}{d\bar{z}} \bar{z}_y. \tag{12.14}$$

Therefore, combining the two previous equations, we have

$$S_{11}(g_x)^2 + 2S_{12}g_xg_y + S_{22}(g_y)^2$$

$$= \pi \left(-\cot \pi(p_1 + p_2) \left(\frac{dg}{dz}(z_x + z_y) + \frac{dg}{d\bar{z}}(\bar{z}_x + \bar{z}_y) \right)^2 \right.$$

$$\left. + \cot \pi p_1 \left(\frac{dg}{dz} z_x + \frac{dg}{d\bar{z}} \bar{z}_x \right)^2 + \cot \pi p_2 \left(\frac{dg}{dz} z_y + \frac{dg}{d\bar{z}} \bar{z}_y \right)^2 \right), \tag{12.15}$$

which is a quadratic form in $\frac{dg}{dz}$ and $\frac{dg}{d\bar{z}}$. We split our further calculation into three claims.

Claim 12.4 *The coefficient of $\left(\frac{dg}{dz} \right)^2$ in (12.15) is 0.*

Proof The coefficient is

$$\pi \left(-(z_x + z_y)^2 \cot \pi(p_1 + p_2) + z_x^2 \cot \pi p_1 + z_y^2 \cot \pi p_2 \right). \tag{12.16}$$

Because $\frac{z_y}{z} - \frac{z_x}{1-z} = 0$ by (12.1), we may write $z_x = \frac{1-z}{z} z_y$ in the previous equation, yielding

$$\pi z_y^2 \left(-\frac{1}{z^2} \cot \pi(p_1 + p_2) + \left(\frac{1-z}{z} \right)^2 \cot \pi p_1 + \cot \pi p_2 \right). \tag{12.17}$$

Substituting

$$z = \frac{\sin \pi p_2}{\sin \pi(p_1 + p_2)} e^{i\pi p_1}, \quad 1 - z = \frac{\sin \pi p_1}{\sin \pi(p_1 + p_2)} e^{-i\pi p_2} \tag{12.18}$$

(the latter of which is easy to see by a similar law-of-sines argument in Figure 12.1) into (12.17) yields

---

[2] Explicit formulas make is straightforward to show that $\det \begin{pmatrix} S_{11} & S_{12} \\ S_{21} & S_{22} \end{pmatrix} = \pi^2$, which actually holds for the surface tensions in greater generality, see (Kenyon and Okounkov, 2006, Section 2.2.3).

$$\frac{\pi z_y^2}{z^2 \sin^2 \pi (p_1+p_2)} \left(-\cos \pi (p_1 + p_2) \sin \pi (p_1 + p_2) + e^{-2\pi i p_2} \cos \pi p_1 \sin \pi p_1 \right.$$
$$\left. + e^{2\pi i p_1} \cos \pi p_2 \sin \pi p_2 \right)$$
$$= \frac{\pi z_y^2}{2z^2 \sin^2 \pi (p_1+p_2)} \left(-\sin 2\pi (p_1 + p_2) + e^{-2\pi i p_2} \sin 2\pi p_1 + e^{2\pi i p_1} \sin 2\pi p_2 \right).$$

Splitting both complex exponentials into real and imaginary parts, the imaginary parts on the second and third of the previous terms cancel, and the real part is

$$- \sin 2\pi (p_1 + p_2) + \sin 2\pi p_1 \cos 2\pi p_2 + \cos 2\pi p_1 \sin 2\pi p_2 = 0$$

by the identity for $\sin(A + B)$. This proves the claim.   □

Claim 12.5   *The coefficient of* $\left(\frac{dg}{d\bar{z}} \right)^2$ *in* (12.15) *is* 0.

Proof   The proof is exactly the same as for Claim 12.4.   □

Claim 12.6   *The coefficient of* $\frac{dg}{dz} \frac{dg}{d\bar{z}}$ *in* (12.15), *multiplied by* $dx\,dy$ *as in the integral in* (12.6), *is equal to* $\frac{dg}{dz} \frac{dg}{d\bar{z}} dz \wedge d\bar{z}$.

Proof   First, because $dz = z_x dx + z_y dy$ and $d\bar{z} = \bar{z}_x dx + \bar{z}_y dy$,

$$dz \wedge d\bar{z} = (z_x \bar{z}_y - \bar{z}_x z_y) dx \wedge dy = \frac{z_y \bar{z}_y}{z \bar{z}} (\bar{z} - z) dx \wedge dy, \qquad (12.19)$$

where, in the last step, we use (12.1) to get

$$z_x \bar{z}_y - \bar{z}_x z_y = z_y \bar{z}_y \left(\frac{1-z}{z} - \frac{1-\bar{z}}{\bar{z}} \right) = \frac{z_y \bar{z}_y}{z \bar{z}} (\bar{z} - z).$$

Thus, it suffices to show the following:

$$\pi (- \cot \pi (p_1 + p_2) \cdot 2(z_x + z_y)(\bar{z}_x + \bar{z}_y) + \cot \pi p_1 \cdot 2 z_x \bar{z}_x + \cot \pi p_2 \cdot 2 z_y \bar{z}_y)$$
$$\overset{?}{=} 2\pi i \frac{z_y \bar{z}_y}{z \bar{z}} (\bar{z} - z). \quad (12.20)$$

Again using $z_x = \frac{1-z}{z} z_y$ to get a common factor of $z_y \bar{z}_y$, and multiplying (12.20) by $z\bar{z}$ and dividing by $2\pi z_y \bar{z}_y$, we see that (12.20) is equivalent to

$$i(\bar{z} - z) \overset{?}{=} -\cot \pi (p_1 + p_2) + (1 - z)(1 - \bar{z}) \cot \pi p_1 + z\bar{z} \cot \pi p_2. \quad (12.21)$$

Substituting (12.18), the left-hand side of (12.21) equals

$$2 \frac{\sin \pi p_1 \sin \pi p_2}{\sin \pi (p_1 + p_2)}, \qquad (12.22)$$

and the right-hand side equals

$$-\cot \pi(p_1 + p_2) + \frac{\sin^2 \pi p_1}{\sin^2 \pi(p_1 + p_2)} \cot \pi p_1 + \frac{\sin^2 \pi p_2}{\sin^2 \pi(p_1 + p_2)} \cot \pi p_2. \quad (12.23)$$

Multiplying through by $\sin^2 \pi(p_1 + p_2)$, we have that (12.20) is equivalent to

$$2 \sin \pi(p_1 + p_2) \sin \pi p_1 \sin \pi p_2 \overset{?}{=} - \cos \pi(p_1 + p_2) \sin \pi(p_1 + p_2)$$
$$+ \sin \pi p_1 \cos \pi p_1 + \sin \pi p_2 \cos \pi p_2.$$

Moving all terms with $\sin \pi(p_1 + p_2)$ to one side, this is the same as

$$\sin \pi(p_1 + p_2)(\cos \pi(p_1 + p_2) + 2 \sin \pi p_1 \sin \pi p_2)$$
$$\overset{?}{=} \sin \pi p_1 \cos \pi p_1 + \sin \pi p_2 \cos \pi p_2.$$

Using the sine double-angle identity and the formula $\cos(A + B)$, the desired identity becomes

$$\sin \pi(p_1 + p_2)(\cos \pi p_1 \cos \pi p_2 - \sin \pi p_1 \sin \pi p_2 + 2 \sin \pi p_1 \sin \pi p_2)$$
$$\overset{?}{=} \frac{1}{2} \sin 2\pi p_1 + \frac{1}{2} \sin 2\pi p_2. \quad (12.24)$$

Because

$$\cos \pi p_1 \cos \pi p_2 - \sin \pi p_1 \sin \pi p_2 + 2 \sin \pi p_1 \sin \pi p_2$$
$$= \cos \pi p_1 \cos \pi p_2 + \sin \pi p_1 \sin \pi p_2 = \cos \pi(p_1 - p_2),$$

(12.24) becomes

$$2 \sin \pi(p_1 + p_2) \cos \pi(p_1 - p_2) \overset{?}{=} \sin 2\pi p_1 + \sin 2\pi p_2. \quad (12.25)$$

This is true by simple trigonometry. We have proved Claim 12.6 and hence Lemma 12.3.   □

                                                                                    □

Exercise 12.7   Repeat the heuristic derivation of the GFF fluctuations for general $c \neq 0$, that is, for the asymptotic of the q^{Volume}-weighted lozenge tilings.

Remark 12.8   One expects that a similar computation (leading to GFF heuristics) can be done for periodically weighted dimers of Kenyon et al. (2006); however, this is not explicitly present in the literature as of 2021. The arguments in Section 3 of Borodin and Toninelli (2018) might turn out to be helpful for simplifying computations for such generalizations.

Remark 12.9   Can we turn the heuristics of this lecture into a rigorous proof of convergence of fluctuations to the GFF? This is very difficult and has not been done. In particular, the very first step (12.2) had an $o(L^2)$ error in the exponent, whereas we would need an $o(1)$ error for a rigorous proof.

13

Lecture 13: Ergodic Gibbs Translation-Invariant Measures

So far, we have discussed the global limits of random tilings. In other words, we were looking at a tiling of a large domain from a large distance, so its rescaled size remained finite. We now start the new topic of *local limits*. The aim is to understand the structure of a random tiling of a large domain near a given point and at the local scale so that we can observe individual lozenges.

We consider a uniformly random tiling of a domain of linear size L and would like to understand the limit as $L \to \infty$ of the probability measure on tilings near $(0,0)$ (see Figure 13.1). There are two basic questions:

1. How do you think mathematically about the limiting object?
2. What are all possible objects (probability measures?) appearing in the limit?

13.1 Tilings of the Plane

As our domain becomes large, its boundary is no longer visible on the local scale, and we deal with lozenge tilings of the whole plane.

Let us identify a tiling with a perfect matching of black and white vertices of the infinite regular hexagonal lattice (see Figure 13.2). Let us denote the set of all perfect matchings with Ω.

By its nature, Ω is a subset of the set of all configurations of edges on the lattice, $\Omega \subset 2^{\text{edges}}$. If we equip 2^{edges} with the product topology and corresponding Borel σ-algebra, then Ω is a measurable subset.

Hence, we can speak about probability measures on Ω, which are the same as probability measures $\mathbb{P}$ on the Borel σ-algebra of 2^{edges}, such that $\mathbb{P}(\Omega) = 1$.

We describe measures on Ω through their *correlation functions*.

Definition 13.1 For $n = 1, 2, \ldots$, the nth correlation function of a probability measure $\mathbb{P}$ on Ω is a function of n edges of the grid:

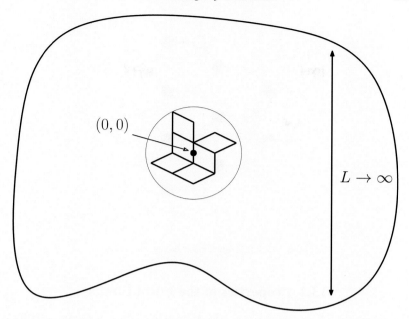

Figure 13.1 Random tiling of a large domain seen locally near the point $(0,0)$.

$$\rho_n(e_1, e_2, \ldots, e_n) = \mathbb{P}(e_1 \in \text{random matching}, \ldots, e_n \in \text{random matching}).$$

Lemma 13.2 *The collection of all correlation functions ρ_n, $n = 1, 2, \ldots$, uniquely determines the measure $\mathbb{P}$.*

Proof By the Caratheodory theorem, any probability measure on 2^{edges} (and hence, on Ω) is uniquely fixed by probabilities of cylinders, which are probabilities of the kind

$\mathbb{P}(\text{a collection of edges} \in \text{matching, another collection of edges} \notin \text{matching}).$

These probabilities can be expressed as finite signed sums of the correlation functions by using the inclusion–exclusion principle. ◻

Exercise 13.3 Consider the following probability measure $\mathbb{P}$: we choose one of the three types of lozenges $\diamondsuit, \oslash, \obslash$ with probability $1/3$ each, and then take the (unique) tiling of the plane consisting of only the lozenges of this type. Compute correlation functions of $\mathbb{P}$.

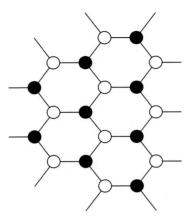

Figure 13.2   Infinite regular hexagonal graph.

13.2 Properties of the Local Limits

When a measure $\mathbb{P}$ on Ω appears as a local limit of uniform measures on tilings of planar domains, it is not an arbitrary measure. Let us list the main properties that we expect such a measure to have.

Definition 13.4   A measure $\mathbb{P}$ on Ω is called *Gibbs* if for any finite subdomain R (= collection of vertices), the conditional measure on the perfect matching of R, given that all vertices of R are matched with vertices of R, but not outside (and in addition, one can condition on any event outside R), is the uniform measure.

The Gibbs property is a way to say that a measure is uniform for the situation when the state space Ω is infinite. Clearly, a uniform measure on the tiling of any finite domain is Gibbs because the conditionings do not change the uniformity. Hence, we expect every local limit of a tiling also to be Gibbs.

Definition 13.5   A measure $\mathbb{P}$ on Ω is called "translation-invariant" if the correlation functions are unchanged upon translations of all arguments by the same vector.

Why do we expect local limits to be translation-invariant? This property should be treated as some kind of a continuity statement. In general, the local limit might depend on the macroscopic zoom-in position inside the domain. However, if we believe that this dependence is continuous, then a shift by a finite vector should become negligible in the large domain limit and, hence, should not change the local limit.

Let us take two Gibbs translation-invariant measures μ_1 and μ_2. For every $0 < \alpha < 1$, the (convex) linear combination $\alpha\mu_1 + (1 - \alpha)\mu_2$ is also a Gibbs translation-invariant measure. Hence, such measures form a convex set.

Definition 13.6 A Gibbs translation-invariant measure is called "ergodic" if it is an extreme point of the set of all such measures, that is, if it cannot be decomposed into a convex linear combination of two distinct measures from this class.

We use the abbreviation *EGTI* for ergodic Gibbs translation-invariant measures.

Lemma 13.7 *If $\mathbb{P}$ is an EGTI probability measure, and A is a translation-invariant event, then either $\mathbb{P}(A) = 0$ or $\mathbb{P}(A) = 1$.*

Remark 13.8 By definition, a translation-invariant event in Ω is a measurable subset of Ω, which is preserved under shifts of the entire grid. For example, if R_n is a growing sequence of convex domains (say, rectangles) exhausting the plane, then

$$\left\{ \limsup_{n\to\infty} \frac{\text{number of horizontal edges of matching inside } R_n}{\text{area of } R_n} \leq \frac{1}{7} \right\}$$

is a translation-invariant event.

Proof of Lemma 13.7 We argue by contradiction. Assume that $0 < \mathbb{P}(A) < 1$, and let $\bar{A}$ be the complementary event. Define $\mathbb{P}_1$ to be the normalized restriction of $\mathbb{P}$ onto the set A; that is:

$$\mathbb{P}_1(B) = \frac{\mathbb{P}(A \cap B)}{\mathbb{P}(A)}.$$

Also define $\mathbb{P}_2$ to be the normalized restriction of $\mathbb{P}$ onto the set $\bar{A}$; that is:

$$\mathbb{P}_2(B) = \frac{\mathbb{P}(\bar{A} \cap B)}{\mathbb{P}(\bar{A})}.$$

Then both $\mathbb{P}_1$ and $\mathbb{P}_2$ are Gibbs translation-invariant measures, and

$$\mathbb{P}(\cdot) = \mathbb{P}(A)\mathbb{P}_1(\cdot) + \mathbb{P}(\bar{A})\mathbb{P}_2(\cdot).$$

Therefore, $\mathbb{P}$ is not ergodic. $\qquad\qquad\square$

Exercise 13.9 Show that the measure $\mathbb{P}$ of Exercise 13.3 is Gibbs and translation-invariant but not ergodic. Represent it as a convex linear combination of EGTI measures.

13.3 Slope of EGTI Measure

Let us introduce an important numeric characteristic of an EGTI measure.

Definition 13.10   The *slope* of an EGTI probability measure on Ω is a triplet $(p_\diamond, p_\square, p_\backslash)$ of nonnegative numbers summing up to 1, defined in either of two equivalent ways:

1. For an arbitrary triangle of the grid (= vertex of the hexagonal lattice), $p_\diamond, p_\square, p_\backslash$ are the probabilities that this triangle belongs to a lozenge of a corresponding type (is matched to one of the three corresponding adjacent vertices of another color).

2. If R_n is a growing sequence of convex domains (say, rectangles) exhausting the plane, then

$$p_\diamond = \lim_{n \to \infty} \frac{\text{number of triangles inside } R_n \text{ covered by lozenges of type } \diamond}{\text{number of triangles inside } R_n},$$

(13.1)

and similarly for $p_\square$ and $p_\backslash$.

Remark 13.11   The probabilities in the first definition of the slope do not depend on the choice of the triangle because of the translation invariance of the measure.

The limits in (13.1) exist almost surely, which is a property that needs to be proven. (If we know that the limits exist, they have to be deterministic because of ergodicity.) One way to prove that the limits in (13.1) exist in probability (rather than the stronger condition of being almost sure limits) is by bounding the variance of the right-hand side, using the explicit description of the EGTI measures given later in this lecture.

As soon as we know that a deterministic limit in (13.1) exists, it has to be equal to $p_\diamond$ from the first definition because this is the expectation of the right-hand side in (13.1).

The possible slopes $(p_\diamond, p_\square, p_\backslash)$ form a triangle: $p_\diamond > 0$, $p_\square > 0$, $p_\backslash > 0$, $p_\diamond + p_\square + p_\backslash = 1$. We say that the slope is extreme if one of the proportions of lozenges, $p_\diamond$, $p_\square$, or $p_\backslash$, is zero, and nonextreme otherwise.

Theorem 13.12 (Sheffield, 2005)   *For each nonextreme slope, there exists a unique EGTI probability measure of such slope.*

Idea of the Proof   Suppose that there are two different EGTI measures of the same slope, and let π_1, π_2 be corresponding random perfect matchings from Ω sampled independently. Superimpose π_1 and π_2 (i.e., consider the union of all edges of the hexagonal lattice appearing in either of the matchings).

Then $\pi_1 \cup \pi_2$ is a collection of double edges (if the same edge belonged to both π_1 and π_2), finite cycles (of alternating edges from π_1 and π_2), and infinite curves extending to infinity (again, of alternating edges from π_1 and π_2).

Claim. Almost surely, $\pi_1 \cup \pi_2$ has double edges and finite cycles but no infinite curves.

The claim is a difficult statement, and we refer the reader to Sheffield (2005) for the proof. Note that for measures of different slopes, it no longer holds.

Given the claim, let us show that the distributions of π_1 and π_2 coincide. For that, we demonstrate a sampling procedure, which is the same for both measures.

Note that for a given a cycle of $\pi_1 \cup \pi_2$, half of its edges belong to π_1, and the other half belongs to π_2. There are two options: either all even edges belong to π_1 (and all odd edges belong to π_2) or all even edges belong to π_2. Applying Gibbs property to this cycle separately for π_1 and π_2, we see that these two options arise with equal probability $\frac{1}{2}$. Hence, we can sample π_1 by the following procedure: first sample $\pi_1 \cup \pi_2$, and then flip an independent coin for each cycle to decide which of its edges belong to π_1. Because the procedure for π_2 is the same, we are done. □

Theorem 13.12 has an interesting corollary. First, it explains why we expect local limits of random tilings to be ergodic: indeed, we want them to have fixed slopes matching the slope of the global macroscopic limit shape. This argument can also be reversed: if we knew a priori that the local limits of random tilings should be given by EGTI measures, then Theorem 13.12 would uniquely fix such a measure. Unfortunately, in practice, checking the EGTI property is hard, with perhaps the most complicated part being translation invariance.

Nevertheless, it was conjectured in Conjecture 13.5 of Cohn et al. (2001) that all the limits are EGTI, which was recently proven in Aggarwal (2019).

Theorem 13.13 (Aggarwal, 2019) *Locally, near any point of a (proportionally growing) domain with a nonextreme slope of the limit shape, the uniform measure on tilings converges to the EGTI measure of the same slope.*

In Lecture 17, we will present a partial result in the direction of Theorem 13.13 from Gorin (2017) after some preparations in Lectures 14, 15, and 16. The proof of the general case in Aggarwal (2019) starts with the same approach and then adds to it several additional ideas that we are not going to present in these lectures.

13.4 Correlation Functions of EGTI Measures

Theorem 13.13 provides a simple way to identify the EGTI measures for each slope. We need to take *any* domain where such slope appears and compute the local limit. In fact, we already made such a computation when analyzing tilings on the torus in Lecture 4, and the result is given by Theorem 4.11. Let us recast the result here.

Corollary 13.14 *For an EGTI measure of slope $(p_\diamond, p_\triangle, p_\triangledown)$, the nth correlation function computing the probability of observing the edges $(x_1, y_1, \tilde{x}_1, \tilde{y}_1), \ldots, (x_n, y_n, \tilde{x}_n, \tilde{y}_n)$ between white vertices (x_i, y_i) and black vertices $(\tilde{x}_i, \tilde{y}_i)$ in a random perfect matching is*

$$\rho_n((x_1, y_1, \tilde{x}_1, \tilde{y}_1), \ldots, (x_n, y_n, \tilde{x}_n, \tilde{y}_n))$$

$$= \prod_{i=1}^{n} K_{00}(x_i, y_i, \tilde{x}_i, \tilde{y}_i) \det_{1 \leq i,j, \leq n} (\tilde{K}^{\alpha,\beta}[\tilde{x}_i - x_j, \tilde{y}_i - y_j]), \quad (13.2)$$

where

$$\tilde{K}^{\alpha,\beta}(x, y) = \oiint_{|z|=|w|=1} \frac{w^x z^{-y}}{1 + e^\alpha z + e^\beta w} \frac{dw}{2\pi i w} \frac{dz}{2\pi i z},$$

where we use the coordinate system of Figure 3.2, $K_{00}(x_i, y_i, \tilde{x}_i, \tilde{y}_i)$ are weights from that figure, and the correspondence between $(p_\diamond, p_\triangle, p_\triangledown)$ and (α, β) is as in Theorem 6.1.

Remark 13.15 At this point, the Gibbs property of the measure might seem to be very far away from the determinantal formulas for the correlation functions. However, the Gibbsianity can actually be directly seen from (13.2), as explained in Borodin and Shlosman (2010).

Let us get comfortable with formula (13.2) by computing the correlation functions of horizontal lozenges $\diamond$ along a vertical line. This corresponds to taking $\tilde{x}_i = 1$, $x_i = 0$ and $y_i = \tilde{y}_i$. Hence, absorbing the $K_{0,0}$ prefactor (which is e^β in this case) into the determinant, we conclude that the correlation functions of horizontal lozenges $\diamond$ at positions y_i are minors of the matrix

$$(y_i, y_j) \mapsto \oiint_{|z|=|w|=1} \frac{e^\beta w z^{y_i - y_j}}{1 + e^\alpha z + e^\beta w} \frac{dw}{2\pi i w} \frac{dz}{2\pi i z}. \quad (13.3)$$

We further argue as in Section 6.1, and compute the w-integral as a residue at the unique pole of the integrand to get

$$\frac{1}{2\pi i} \int_{\bar{z}_0}^{z_0} z^{y_i - y_j - 1} dz = \frac{z_0^{y_i - y_j} - \bar{z}_0^{y_i - y_j}}{2\pi i (y_i - y_j)} = \frac{|z_0|^{y_i - y_j}}{\pi (y_i - y_j)} \sin(\mathrm{Arg}(z_0)(y_i - y_j)).$$

Note that the $|z_0|^{y_i - y_j}$ prefactor cancels out when we compute the minors. By looking at the $y_i - y_j = 0$ case, we also match $\text{Arg}(z_0)$ with $\pi p_\diamond$. Hence, we reach a conclusion.

Corollary 13.16   *For the EGTI probability measure of slope* $(p_\diamond, p_\lozenge, p_\llcorner)$

$$\mathbb{P}(\diamond \text{ at positions } (0, y_1), \ldots, (0, y_n)) = \det[K(y_i - y_j)]_{i,j=1}^n,$$

with

$$K(y) = \begin{cases} \dfrac{\sin(\pi p_\diamond y)}{\pi y}, & y \neq 0, \\[2mm] p_\diamond, & y = 0. \end{cases} \tag{13.4}$$

Remark 13.17   Rotating by 120 degrees, we can also get similar statements for the $\lozenge$ and $\llcorner$ lozenges – their distributions along appropriate sections will be described by (13.4), with parameter $p_\diamond$ replaced by $p_\lozenge$ and $p_\llcorner$, respectively.

Exercise 13.18   Perform the computation of Remark 13.17 in detail.

The kernel of (13.4) is called the "discrete sine kernel," and the point process of Corollary 13.16 is called the "discrete sine process."

The continuous counterpart of the sine process (when y's become real numbers rather than integers) is a *universal object* in the random matrix theory,[1] where it describes local limits for the eigenvalues in the bulk of the spectrum of large complex Hermitian matrices; see Kuiklaars (2011), Tao and Vu (2012), Lubinsky (2016), and Erdos and Yau (2017) for recent reviews. Conjecturally, the sine process also appears in many other settings, such as large eigenvalues of the Laplace operator in various domains or even spacings between the zeros of the Riemann zeta-function.

13.5 Frozen, Liquid, and Gas phases

An important property of the EGTI measure with a nondegenerate slope $(p_\diamond, p_\lozenge, p_\llcorner)$ satisfying $p_\diamond, p_\lozenge, p_\llcorner > 0$ is the polynomial decay of the correlations between lozenges with distance. This can be directly seen from the result of Corollary 13.14 because $\tilde{K}^{\alpha,\beta}(x, y)$ decays polynomially in x and y. This type of EGTI measure is called the "liquid phase" in Kenyon et al. (2006), with an alternative name being the "rough unfrozen phase."

---

[1]   We return to the connection between tilings and random matrix theory in Lectures 19 and 20.

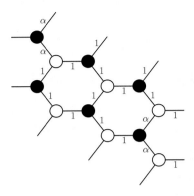

Figure 13.3 The 3×2 periodic weights used in Figure 13.4 with $\alpha = 10$.

Extreme slopes (when either $p_\diamond$, or $p_\square$, or $p_\searrow$ vanishes) lead to very different structures of the EGTI measure: the random tiling becomes degenerate and has nondecaying correlations. This is the *frozen phase*.

For uniformly random lozenge tilings, the liquid and frozen phases exhaust the list of possible types of EGTI measures appearing in local limits. However, we can allow more general nonuniform weights of tilings. One natural choice of nonuniformity is by declaring that whenever a cube is being added to the stepped surface encoding tiling – that is, whenever one tiling differs from another one by an increase of 3 of the value of the height function (in the terminology of Lecture 1) at a single point (x, y) – the weight of the tiling is multiplied by $w(x, y)$. If we require $w(x, y)$ to be a *periodic* function of x and y, then we arrive at *periodically weighted* lozenge tilings. In particular, if the period is 1 and $w(x, y) = q$, then these are the q^{Volume}-weighted tilings, which we have already encountered in the previous lectures. A simpler but slightly less general way to introduce periodic weighting is by putting the weights on edges of the hexagonal grid for perfect matchings, as we did in Lectures 3 and 4 (cf. Figure 3.2). The second approach does not allow one to obtain q^{Volume} weighting with $q \neq 1$, but if we ignore this fact, the two approaches are equivalent.

Kenyon et al. (2006) produced a complete classification of possible EGTI measures for periodically weighted tilings. In particular, they have shown that for certain choices of weights, a third type of EGTI measure appears. They called these new measures the "gas phase," with an alternative name being the "smooth unfrozen phase." In this phase, the correlations between lozenges decay *exponentially*, and the height function is almost linear, with small isolated defects. Kenyon et al. (2006) constructed EGTI measures by a limit transition from the torus, similarly to what we did in Lecture 4. Obtaining the gas phase as

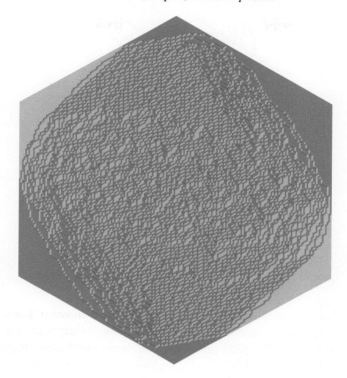

Figure 13.4 A 3×2 periodic random lozenge tiling of a $100 \times 100 \times 100$ hexagon. Two gas phases are visible adjacent to the top and bottom green frozen regions. Let us note that, asymptotically, a gas phase cannot be adjacent to a frozen phase; this would contradict Theorem 4.1 in De Silva and Savin (2010), which says that discontinuity of lozenge densities can happen only at boundary slopes. Hence, there is, in fact, a very narrow liquid region separating the gas and frozen phases. (Source: Simulation courtesy of Christophe Charlier and Maurice Duits.)

a local limit of tilings of planar polygonal domains is a more complicated task. It was first achieved for the 2×2 periodic domino tilings of the Aztec diamond in Chhita and Young (2014); Chhita and Johansson (2016). A systematic approach to periodic situations was introduced in Duits and Kuijlaars (2019), and there is a hope that the gas phase in periodically weighted lozenge tilings of polygons can be rigorously studied using this approach; yet, it was not done at the time when this book was finished.

Figure 13.4 shows a simulation[2] of the 3×2 periodically weighted lozenge tilings of the hexagon with weights assigned to edges in a periodic way with the

---

[2] It was produced using an adaptation of the algorithm from Propp (2003).

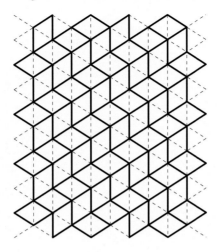

Figure 13.5   A staircase-like frozen phase with two types of lozenges.

fundamental domain shown in Figure 13.3; see also the web page of Borodin and Borodin[3] for a three-dimensional virtual reality version of the same simulation. We note that the results of Kenyon et al. (2006) imply that 3×2 is the minimal size of the fundamental domain, which makes the appearance of a gas phase possible.

The phenomenology for periodically weighted tilings becomes richer than for the uniform case, which we mostly study in this book, and we refer the reader to Mkrtchyan (2019), Charlier et al. (2020), Berggren and Duits (2019), Charlier (2020), and Beffara et al. (2020) for some results. Just as an example, whereas for the uniformly random lozenge tilings we could observe only three frozen phases corresponding to 3 types of lozenges, for the 2×2 periodic case, another frozen phase is possible, in which the densities of two types of lozenges are equal to $1/2$, and the corresponding stepped surface looks like a staircase; see Figure 13.5.

---

[3]   A. Borodin and M. Borodin, A 3D representation for lozenge tilings of a hexagon, http://math.mit.edu/~borodin/hexagon.html.

14

Lecture 14: Inverse Kasteleyn Matrix for Trapezoids

Our goal over the next few lectures is to discuss a local limit theorem for the trapezoid (or sawtooth) domain of Figure 14.1. It is so named because of the "dents" along its right edge. We are going to prove that the local structure of uniformly random lozenge tilings for such domains is asymptotically governed by the ergodic Gibbs translation-invariant (EGTI) measures introduced in the previous lecture.

Why should we study this particular class of domains?

- The sawtooth domains are *integrable* in the sense that they have an explicit inverse Kasteleyn matrix.

- This type of domain has links to representation theory. In particular, consider a sawtooth domain with width N and dents at locations $t_1 > \cdots > t_N$. Writing $\lambda_j = t_j - (N - j)$, we see that $\lambda_1 \geq \cdots \geq \lambda_N$. Each irreducible representation of the unitary group $U(N)$ has a *signature*, and it turns out that the number of tilings of the sawtooth domain is given by the (vector space) dimension of the representation having signature $(\lambda_1, \ldots, \lambda_N)$, and by the Weyl dimension formula, this is given by $\prod_{i<j} \frac{(\lambda_i - i) - (\lambda_j - j)}{j - i}$. We come back to this computation in Lecture 19; see Proposition 19.3.

- The boundary of every domain locally looks like a trapezoid; eventually, the results for sawtooth domains and their extensions can be used as a building block for proving the local limit theorem for general domains; see Aggarwal (2019) and Gorin (2017).

In Lecture 3 (see Theorem 3.5), we expressed the correlation functions of random tilings through the inverse Kasteleyn matrix. The goal of this lecture is to compute this inverse matrix explicitly, which will be the main tool for our analysis in the next lecture.

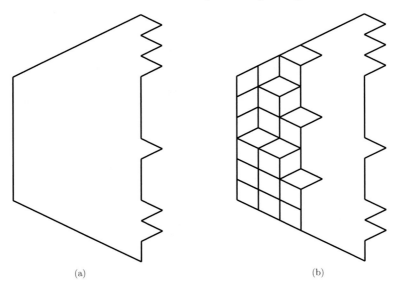

(a)   (b)

Figure 14.1 **Panel (a)**: A trapezoid/sawtooth domain. **Panel (b):** Observe
that if the width of the domain is N, then the domain is tileable by lozenges if
and only if its right border has N dents. Indeed, it is clear by inspection that
the first layer has exactly one horizontal lozenge, the second has two, and so
forth.

Following the notations of Petrov (2014a) (which this lecture is based
on), we apply a rotation and skew transformation to the sawtooth domain,
as in Figure 14.2. This is helpful because the triangles of the grid are now
parameretized by pairs of integers; the correspondence is shown in panel (b) of
Figure 14.2.

The Kasteleyn (adjacency) matrix of the dual graph of triangles, restricted
to the sawtooth domain, is as follows: if $\triangledown(x, n)$ and $\triangle(y, m)$ are both inside
the domain, then

$$K(\triangledown(x, n); \triangle(y, m)) = \begin{cases} 1, & \text{if } (y, m) = (x, n), \\ 1, & \text{if } (y, m) = (x, n - 1), \\ 1, & \text{if } (y, m) = (x + 1, n - 1), \\ 0, & \text{otherwise.} \end{cases}$$

If one of the triangles $\triangledown(x, n)$ or $\triangle(y, m)$ is outside the domain, then
$K(\triangledown(x, n); \triangle(y, m)$ vanishes.

The following theorem is the main result of this lecture. We use the
Pochhammer symbol $(u)_a = u(u + 1) \cdots (u + a - 1)$.

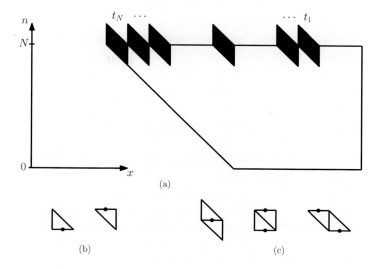

Figure 14.2 **Panel (a):** After rotating and shearing the diagram, all lattice lines are either grid aligned or at a 45-degree angle. We set coordinates (x, n), where x is the horizontal coordinate, and n is the vertical coordinate. We only care about x up to additive constant, so the position of the vertical axis can be chosen arbitrarily. n is set to be 0 at the bottom boundary and N at the top boundary. We understand the domain to be the sawtooth region with its N dents removed (white portion of figure). **Panel (b):** There are now two types of lattice triangles. We place a root on the horizontal edge of each triangle and define the coordinate of the triangle to be the coordinate of its root. **Panel (c):** There are three types of lozenges we can make out of triangles $\triangledown(x, n)$ and $\triangle(y, m)$. From left to right: $(y, m) = (x, n)$, $(y, m) = (x, n - 1)$, $(y, m) = (x + 1, n - 1)$.

Theorem 14.1 (Petrov, 2014a, Theorem 1)   *As in Figure 14.2, consider a sawtooth domain with height N and dents at $(t_1, N), \ldots, (t_N, N)$. The inverse Kasteleyn matrix $K^{-1}(\triangle(y, m); \triangledown(x, n))$ is given by*

$$K^{-1}(\triangle(y, m); \triangledown(x, n)) = (-1)^{y-x+m-n+1} \mathbb{1}_{m<n} \mathbb{1}_{y \leq x} \frac{(x - y + 1)_{n-m-1}}{(n - m - 1)!}$$

(14.1)

$$+ \frac{(-1)^{y-x+m-n}}{(2\pi i)^2} \frac{(N - n)!}{(N - m - 1)!} \oint_{\{y, \ldots, t_1\}} \oint_{\infty} \frac{dw \, dz}{w - z} \frac{(z - y + 1)_{N-m-1}}{(w - x)_{N-n+1}} \prod_{r=1}^{N} \frac{w - t_r}{z - t_r},$$

(14.2)

where the z-contour, C_z, is a loop around $\{y, y + 1, \ldots, t_1\}$ (and enclosing no poles of the integrand outside this set), and the w-contour, C_w, is a large loop around C_z and all poles of the integrand.

Remark 14.2   We define our domain to be the trapezoid with dents *removed*, and therefore, the previous formula is applicable only for triangles $\triangle(y,m), \triangledown(x,n)$, which lie entirely in the white region in Panel (a) of Figure 14.2. In particular, we do not allow $\triangledown(x,n) \in \{\triangledown(t_1, N), \ldots, \triangledown(t_N, N)\}$ because these correspond to the dents.

Remark 14.3   Our proof of Theorem 14.1 is through *verification*, and before proceeding to it, let us discuss how the statement can be obtained.

Historically, determinantal formulas for the *two-dimensional* correlation functions were first deduced by Eynard and Mehta (1998) in the study of coupled random matrices, with the particular case given by the Dyson Brownian motion. The latter is the evolution of the eigenvalues of a Hermitian matrix, whose matrix elements evolve in time as independent Brownian motions. Another equivalent definition identifies the Dyson Brownian motion with N independent Brownian motions conditioned to have no collisions, which makes the analogy with tilings in the form of Section 2.2 clear.

Simultaneously, Brézin and Hikami (1996, 1997) introduced double contour integral formulas for the correlation kernel for the fixed time distribution of the Dyson Brownian motion started from an arbitrary initial condition.

During the next 15 years, double contour integrals were found in numerous settings of random matrices and random tilings. The closest to the context of Theorem 14.1 instances are the formulas for the random plane partitions in Okounkov and Reshetikhin (2003)[1] and for the continuous Gelfand–Tsetlin patterns in Metcalfe (2013).

Thus, the appearance of the double contour integrals in the setting of Theorem 14.1 was not completely unexpected. The arguments in Petrov (2014a) (see also Duse and Metcalfe, 2015) use the Eynard–Mehta theorem to produce determinantal formulas and then introduce a nice trick by inverting a matrix, which enters into the correlation kernel, by using contour integrals.

Proof of Theorem 14.1   Throughout the proof, by $K^{-1}(\triangle(y,m); \triangledown(x,n))$, we mean the expression of (14.1), (14.2). We would like to evaluate $K^{-1}K(\triangle(y,m); \triangle(y',m'))$ for each pair $(y,m), (y',m')$, and show that it is the delta function $\delta(y = y'; m = m')$.

Case 1: Triangle $\triangle(y',m')$ has all its neighbors $\triangledown$ inside the domain.
  We need to show that

$$K^{-1}(\triangle(y,m); \triangledown(y',m')) + K^{-1}(\triangle(y,m); \triangledown(y',m'+1))$$
$$+ K^{-1}(\triangle(y,m); \triangledown(y'-1,m'+1)) \stackrel{?}{=} \delta(y=y'; m=m'). \quad (14.3)$$

[1] We will return to this model in Lectures 22–23.

We first compute the double integral term (14.2) for the left-hand side of (14.3) and verify that it is zero. Indeed, the integrand is a multiple of

$$\frac{(N-m')!}{(w-y')_{N-m'+1}} - \frac{(N-m'-1)!}{(w-y')_{N-m'}} + \frac{(N-m'-1)!}{(w-y'+1)_{N-m'}}$$

$$= \frac{(N-m'-1)!}{(w-y')_{N-m'+1}} \left((N-m') - (w-y'+N-m') + (w-y') \right) = 0.$$

Next, we check that the first term (14.1) after summation in the left-hand side of (14.3) gives $\delta(y=y'; m=m')$. We have four subcases:

- $(y,m) = (y',m')$. Then a direct evaluation yields $-0 + \frac{(1)_0}{0!} - 0 = 1$, as needed.
- $m' < m$. All the indicators are zero, so this evaluates to zero.
- $m' = m$, $y' \neq y$. Then the sum reduces to $0 + \mathbb{1}_{y \le y'} - \mathbb{1}_{y \le y'}$, which is zero.
- $m' > m$. In this case, $\mathbb{1}_{m<m'} = 1$ for all terms, and we want to verify that

$$\mathbb{1}_{y \le y'} \frac{(y'-y+1)_{m'-m-1}}{(m'-m-1)!} - \mathbb{1}_{y \le y'} \frac{(y'-y+1)_{m'-m}}{(m'-m)!} + \mathbb{1}_{y \le y'-1} \frac{(y'-y)_{m'-m}}{(m'-m)!} = 0.$$

  - When $y > y'$, all terms are zero because the indicators are zero.
  - When $y = y'$, we indeed get

$$\frac{(1)_{m'-m-1}}{(m'-m-1)!} - \frac{(1)_{m'-m}}{(m'-m)!} + 0 = 0,$$

   because $(1)_a = a!$.
  - When $y < y'$, we have

$$\frac{(y'-y+1)_{m'-m-1}}{(m'-m)!} \left((m'-m) - (y'-y+m'-m) + (y'-y) \right) = 0,$$

   as needed.

This finishes the proof for the case when triangle $\triangle(y', m')$ has all its neighbors $\triangledown$ inside the domain. Note that the situations when $\triangle(y', m')$ is on a vertical boundary of the trapezoid, or on its top boundary, but not adjacent to dents, are included in this case.

Case 2: $\triangle(y', m')$ lies on the bottom boundary (i.e., $m' = 0$).

In this case, we claim that $K^{-1}((y,m);(y',m')) = 0$, so the analysis of the previous case is sufficient. The first term (14.1) is zero because $\mathbb{1}_{m<m'} = 0$. When $w \to \infty$, the w-integrand of the contour integral (14.2) decays as $\frac{1}{|w|^2}$, and so taking the contour C_w to be a circle of large radius R, we see that (14.2) has magnitude bounded by R^{-1} and hence must vanish.

Case 3: $\triangle(y', m')$ is on the top boundary and is adjacent to one or two dents $\triangledown$.

As in the case of the bottom boundary, it suffices to show that if there is a dent $\triangledown(y' - 1, m' + 1)$ (respectively, $\triangledown(y', m' + 1)$), then $K^{-1}(\triangle(y, m); \triangledown(y' - 1, m' + 1)) = 0$ (respectively, $K^{-1}(\triangle(y, m); \triangledown(y', m' + 1)) = 0$). We only prove the former; the proof for the latter is similar.

Consider the double contour integral term (14.2) of $K^{-1}(\triangle(y, m); \triangledown(y' - 1, m' + 1))$. Observe that in our case, because $\triangledown(y' - 1, m' + 1)$ is a dent, we have $y' - 1 = t_r$ for some r and $m' = N - 1$. Hence, the factor in the denominator $(w - (y' - 1))_{N-(m'+1)+1} = w - (y' - 1)$ cancels with the factor $(w - t_r)$ in the numerator of the integrand. Therefore, the only pole of the w-integrand is at $w = z$ arising from $\frac{1}{w-z}$. Hence, the w-integral evaluates as the residue at $w = z$, which is computed in Lemma 14.4 to be $\mathbb{1}_{y \leq y'} \frac{(m'-m)_{y'-y}}{(y'-y)!}$. Because $\frac{(y'-y+1)_{m'-m-1}}{(m'-m-1)!} = \frac{(m'-m)_{y'-y}}{(y'-y)!}$, we conclude that the double integral (14.2) is exactly the negative of the indicator term (14.1), so that $K^{-1}(\triangle(y, m); \triangledown(y' - 1, m' + 1)) = 0$ as needed. We are done with this case and hence with the proof of Theorem 14.1. □

Lemma 14.4 *Take integers y, y', m, m' and assume $m' \geq m$. Then,*

$$\frac{(N - m')!}{(N - m - 1)!} \frac{1}{2\pi i} \oint_{y,y+1,y+2,\ldots} \frac{(z - y + 1)_{N-m-1}}{(z - y')_{N-m'+1}} \, dz = \mathbb{1}_{y \leq y'} \frac{(m' - m)_{y'-y}}{(y' - y)!},$$

where the integration contour encloses all the poles of the integrand to the left from or equal to y.

Proof   We first rephrase the statement to reduce notations. For suitably defined $A, B > 0$ and Δx, the desired statement can be written as

$$\frac{A!}{B!} \frac{1}{2\pi i} \oint_{\widetilde{C}_z} \frac{(z + \Delta x + 1)_B}{(z)_{A+1}} \, dz = \mathbb{1}_{\Delta x \geq 0} \frac{(B - A + 1)_{\Delta x}}{(\Delta x)!}, \qquad (14.4)$$

where the contour $\widetilde{C}_z$ surrounds the integers $-\Delta x, \ldots, 0$. We prove (14.4) by induction. We need the following base cases where (14.4) is easy to check:

- Case $\Delta x < 0$. Then, the right-hand side is clearly zero, and the left-hand side is also zero because there are no singularities enclosed by the contour.
- Case $\Delta x \geq 0$ and $A = B$. Then, the right-hand side is clearly 1, and the integrand of the left-hand side has no poles outside of $\widetilde{C}_z$, so we may enlarge the contour to infinity, and we thus see that the residue is 1 (there's one more linear-in-z factor in the denominator than in the numerator).

We also need the following linear relations: Write $L(A, B, \Delta x)$ and $R(A, B, \Delta x)$ for the left-hand side and right-hand side of (14.4), respectively. Then we have

$$L(A, B + 1, \Delta x) - L(A, B + 1, \Delta x - 1) = L(A, B, \Delta x), \qquad (14.5)$$

$$R(A, B + 1, \Delta x) - R(A, B + 1, \Delta x - 1) = R(A, B, \Delta x). \tag{14.6}$$

Indeed, to prove the first of these relations, we observe that the integrand in the is equal to that in the right-hand side:

$$\left(\frac{z + \Delta x + B + 1}{B + 1} - \frac{z + \Delta x}{B + 1}\right) \frac{A!}{B!} \frac{1}{2\pi i} \frac{(z + \Delta x + 1)_B}{(z)_{A+1}} = \frac{A!}{B!} \frac{1}{2\pi i} \frac{(z + \Delta x + 1)_B}{(z)_{A+1}}.$$

Likewise, we can check the second linear relation (14.6):

$$\left(\frac{B - A + 1 + \Delta x}{B - A + 1} - \frac{\Delta x}{B - A + 1}\right) \mathbb{1}_{\Delta x \geq 0} \frac{(B - A + 1)_{\Delta x}}{(\Delta x)!} = \mathbb{1}_{\Delta x \geq 0} \frac{(B - A + 1)_{\Delta x}}{(\Delta x)!}.$$

Consequently, if we know that (14.4) holds for two of the three triples $(A, B + 1, \Delta x)$, $(A, B + 1, \Delta x - 1)$, $(A, B, \Delta x)$, then it also holds for the third one.

We now prove (14.4) by induction on Δx. We know that (14.4) holds for all triples $(A, B, \Delta x)$ with $\Delta x < 0$. Now suppose that for some $\Delta x \geq 0$, we know for all A, B that (14.4) holds for $(A, B, \Delta x - 1)$. For any A, we also know that (14.4) holds for the triple $(A, A, \Delta x)$. Repeatedly using (14.5) and (14.6), we extend from $B = A$ to the arbitrary B case and conclude that (14.4) holds for all $(A, B, \Delta x)$. □

Exercise 14.5 Consider a macroscopic rhombus or, equivalently, an $A \times B \times 0$ hexagon (which is a particular case of a trapezoid). It has a unique lozenge tiling, and therefore the correlation functions for the uniformly random tilings of this domain are straightforward to compute. Check that they match the expression given by the combination of Theorems 3.5 and 14.1.

Lecture 15: Steepest Descent Method for Asymptotic Analysis

15.1 Setting for Steepest Descent

In this lecture we discuss how to understand the asymptotic behavior of the integrals similar to that of the correlation function for trapezoids in Theorem 14.1. A general form of the integral is

$$\oint_{C_w} \oint_{C_z} \exp(N(G(z) - G(w)))\, f(z, w)\, \frac{dz\, dw}{z - w}. \qquad (15.1)$$

This sort of integral arises frequently in integrable probability, and as such, there is a well-developed machinery to deal with it. The key features of (15.1) are as follows:

- The integrand grows exponentially with N. In (14.2), this is due to the $\prod_{r=1}^{N}$ product and growing Pochhammer symbols.
- The leading contributions in z and w are inverse to each other, hence $G(z) - G(w)$ in the exponent of (15.1).
- The integrand has a simple pole at $z = w$.
- Additional factors in the integrand denoted $f(z, w)$ grow subexponentially in N.

We start with simpler examples and work our way toward (15.1).

15.2 Warm-Up Example: Real Integral

Let us compute the large N asymptotics of the real integral

$$\int_0^1 \exp(N(x - x^2))\, dx.$$

We use the Laplace method. Write $f(x) = x - x^2$; this is a concave function on $[0, 1]$ with a maximum at $\frac{1}{2}$. The integral is dominated by a neighborhood of $\frac{1}{2}$ (indeed, $\exp(Nf(x))$ is exponentially smaller away from $\frac{1}{2}$), so we do a Taylor expansion there:

$$f(x) = \frac{1}{4} - \left(x - \frac{1}{2}\right)^2.$$

Making the substitution $y = \sqrt{N}(x - \frac{1}{2})$, we get

$$\int_0^1 \exp\left(N\left(\frac{1}{4} - \left(x - \frac{1}{2}\right)^2\right)\right) dx = e^{\frac{N}{4}} \frac{1}{\sqrt{N}} \int_{-\sqrt{N}/2}^{\sqrt{N}/2} e^{-y^2} dy$$

$$= e^{\frac{N}{4}} \frac{1}{\sqrt{N}} \int_{-\infty}^{\infty} e^{-y^2} dy \cdot (1 + o(1)) = e^{\frac{N}{4}} \sqrt{\frac{\pi}{N}} \cdot (1 + o(1)).$$

The conclusion is that the integral is dominated by the neighborhood of the point where $f(x)$ is maximized. This point can be found by solving the critical point equation $f'(x) = 0$.

Exercise 15.1 Prove Stirling's formula by finding the large N asymptotics of the integral

$$N! = \int_0^\infty x^N \exp(-x)\, dx.$$

(Hint: Start by changing the variables $x = Ny$.)

15.3 One-Dimensional Contour Integrals

For the next step, we take $K = \alpha N$ and study for large N the binomial coefficient

$$\binom{N}{K} = \frac{1}{2\pi i} \oint \frac{(1 + z)^N}{z^{K+1}}\, dz,$$

where the contour encloses the origin. (Because $\frac{(1+z)^N}{z^{K+1}}$ has only a single pole at $z = 0$, any contour surrounding 0 gives the same result.)

Inspired by the previous case, we might think that the integral is dominated by a neighborhood of the maximizer of $|1 + z|^N / |z|^K$ on the integration contour. This turns out to be false; indeed, if this worked, one could vary the contour and get a different answer! The problem is that there could be huge oscillations in the complex argument (i.e., angle) of the integrand, causing cancellations. We generalize the previous approach in a different way.

We write $f(z) = \ln(1 + z) - \alpha \ln z$ so that the integral can be written as

$$\frac{1}{2\pi i} \oint \exp(Nf(z)) \, dz.$$

Notice that

$$f'(z) = \frac{1}{1 + z} - \frac{\alpha}{z}.$$

Hence, f has a unique critical point

$$z_c = \frac{\alpha}{1 - \alpha},$$

and

$$f''(z_c) = \frac{(1 - \alpha)^3}{\alpha} > 0.$$

f admits the Taylor expansion in a neighborhood of z_c:

$$f(z) = f(z_c) + \frac{1}{2}f''(z_c)(z - z_c)^2 + \cdots.$$

Notice that along a vertical segment through z_c, because $(z - z_c)^2 < 0$, the real part of $f(z)$ is *maximized* at z_c (it would have been minimized for the horizontal contour). Thus, we want to choose a contour through z_c that is locally vertical at z_c. But how do we specify the contour globally? The trick is to use the level lines of $\Im f$ – imaginary part of the complex function f.

Lemma 15.2   *Take a function $f(z)$ with a critical point z_c such that $f''(z_c) \neq 0$. Suppose that $f(z)$ is holomorphic in a neighborhood of z_c. Then there are two curves γ passing transversally to each other through z_c and such that $\Im f(\gamma)$ is constant along each curve. Along one of these curves γ, the function $\Re f$ decreases as we move away from z_c, until we reach either another critical point of f or a point of nonanalyticity. The same is true for the other curve, except that $\Re f$ instead increases as we move away from z_c.*

Proof   Using Taylor expansions in a neighborhood of z_c, we see four level lines $\Im f(z) = $ const staring at $z = z_c$ in four different directions; they form two smooth curves passing through z_c. Consider the curve for which $\Re f$ decreases locally as we move away from z_c, and suppose that when moving along the curve, we reach a local minimum (along the curve) of $\Re f$ at a point w. If f is not analytic at w, we are done, so assume it is. Because w is a minimizer of $\Re f$ along γ, we have $\frac{d}{d\gamma}|_w \Re f = 0$, but because γ is a level line of $\Im f$, we also have $\frac{d}{d\gamma}|_w \Im f = 0$. Combining these yields $f'(w) = 0$, so w is a critical point of f. The same argument holds for the other curve. $\square$

In our case $f(z) = \ln(1 + z) - \alpha \ln z$, so the function is holomorphic in the slit domain $\mathbb{C}\backslash(-\infty, 0]$. Let's take the locally vertical contour $\Im f(z) = $ const passing through z_c and try to understand its shape. By symmetry, it suffices to understand the half-contour γ starting at z_c and going upward:

- γ cannot escape to infinity. Indeed, $\Re f$ is bounded from above on γ (because it is decreasing away from z_c), but $\exp(N\Re f(z)) = |1 + z|^N/|z|^K$ grows as $|z| \to \infty$.
- Similarly, γ cannot hit 0 because $|1 + z|^N/|z|^K$ explodes near 0.
- Finally, γ cannot hit a point in $\mathbb{R}_+ = (0, +\infty)$. If it did, γ and its conjugate would together form a closed contour in the slit domain $\mathbb{C}\backslash(-\infty, 0)$ along which the harmonic function $\Im f$ is constant. By the maximum principle of harmonic functions, we conclude that $\Im f$ is constant on the whole domain, which is false.

We conclude that γ starts at z_c and exits the upper half-plane somewhere along $\mathbb{R}_- = (-\infty, 0)$. We take for our contour $\gamma \cup \overline{\gamma}$ – we can deform the original contour of integration into $\gamma \cup \overline{\gamma}$ without changing the value of the integral. At this point, by the same analysis as for the warm-up example (the integral is dominated by a neighborhood of z_c because $\exp(Nf(z))$ is exponentially smaller away from z_c), we see that

$$\binom{N}{K} = \frac{(1 + o(1))}{2\pi i}\frac{(1 + z_c)^N}{z_c^{K+1}} \int_{z_c-i\infty}^{z_c+i\infty} \exp\left(\frac{1}{2}N\frac{(1-\alpha)^3}{\alpha}(z - z_c)^2\right) dz$$

$$= (1 + o(1))\frac{(\frac{1}{\alpha})^N}{\left(\frac{1-\alpha}{\alpha}\right)^{K+1}}\frac{1}{\sqrt{2\pi N}}\sqrt{\frac{\alpha}{(1-\alpha)^3}}.$$

Exercise 15.3 Prove the Stirling formula by finding the large N asymptotics of the contour integral around 0:

$$\frac{1}{N!} = \frac{1}{2\pi i}\oint e^z \frac{dz}{z^{N+1}}.$$

15.4 Steepest Descent for a Double Contour Integral

Now we turn to the double contour integral (15.1). A typical situation is that $G(z)$ is real on the real line, which implies $G(\overline{z}) = \overline{G(z)}$. Hence, the solutions to $G'(z) = 0$ come in complex-conjugate pairs. For now, we do not discuss the question of choosing the correct critical point and assume that $G(z)$ has a unique pair of critical points: $z_c, \overline{z}_c$. From these critical points, we construct level lines of $\Im G$ to obtain a pair of contours $\widetilde{C}_w, \widetilde{C}_z$ (see Figure 15.1 for a schematic drawing), different from the original C_w, C_z. Note that near the critical point z_c, two level lines intersect orthogonally; $\Re G$ has a maximum along one of them at z_c and a minimum along another. This is good for us because we would like to have a maximum for both $\Re G(z)$ and $\Re(-G(w))$.

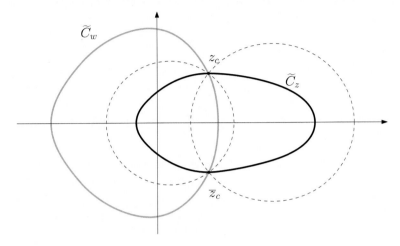

Figure 15.1   A different pair of contours for (15.1). Here, z_c and $\overline{z}_c$ are critical points of G. The dashed curves are the level lines of $\Re G$, and the black and gray contours are the level lines of $\Im G$.

In two dimensions, it turns out that the singularity $\frac{1}{z-w}$ is integrable because $\widetilde{C}_z, \widetilde{C}_w$ are transverse. Thus we can use the methods of the previous case to understand the integral

$$\oint_{\widetilde{C}_w} \oint_{\widetilde{C}_z} f(z,w) \exp(N(G(z) - G(w))) \frac{dz\,dw}{z-w}, \qquad (15.2)$$

by expanding around z_c and $\overline{z}_c$ for the level lines $\widetilde{C}_w, \widetilde{C}_z$. The exponentially large prefactor vanishes (because it is a reciprocal for the w and z integrals), leaving only the polynomially decaying term. Indeed, we only care about (z, w) close to (z_c, w_c); more precisely, we should set $z - z_c = \frac{\tilde{z}}{\sqrt{N}}$ and $w - w_c = \frac{\tilde{w}}{\sqrt{N}}$. After this change of variables, the integrand behaves as

$$\frac{\left(\frac{1}{\sqrt{N}}\right)^2}{\frac{1}{\sqrt{N}}} \to 0.$$

We conclude that this integral goes to zero as $N \to \infty$.

Does this mean the integrals of the kind (15.1) always vanish asymptotically? Note quite, because (15.1) differs from (15.2) by the choice of contours. Hence, we should account for additional residues that might arise in the contour deformation. If we go back to the setting of Theorem 14.1, we see that the original contours are disjoint, whereas the new ones intersect. Hence, in the deformation, we pick up the residue at $z = w$ arising from the $\frac{1}{z-w}$ factor in the integrand. In principle, we might have needed to cross other poles, coming from the remaining factors of the integrand. Yet, let us assume that this is not happening. Then we can conclude the following:

$$\oint_{C_w} \oint_{C_z} f(z, w) \exp(N(G(z) - G(w))) \frac{dz\, dw}{z - w} = \int_{\bar{z}_c}^{z_c} \mathrm{Res}_{z=w}\, dz + O\left(\frac{1}{\sqrt{N}}\right),$$

where the integral is over the part of $\widetilde{C}_z$ between $\bar{z}_c$ and z_c. If the factor $f(z, w)$ in (15.1) is nonsingular, then the residue at $z = w$ is equal to $f(z, z)$, which needs to be integrated from $\bar{z}_c$ to z_c. This integral gives the leading asymptotic of (15.1).

16

Lecture 16: Bulk Local Limits for Tilings
of Hexagons

In this lecture we use the material of the previous few lectures to find the
asymptotic for the correlation functions of uniformly random tilings near a
point of a large hexagonal domain. Specifically, we apply the steepest-descent
techniques of the previous lecture to find the asymptotic of the double contour
integral formula for the entries of the inverse Kasteleyn matrix from Lecture 14.

Recall that we use shifted coordinates, so the hexagonal domain has two
horizontal and two vertical edges. For simplicity, we consider a hexagon with
all side lengths $N/2$. Let us relate the trapezoids (or sawtooth domains) studied
in previous lectures to the hexagons. If we consider a trapezoid with all dents
located in the extreme right-most and left-most positions along the top side
(shown as solid black lozenges in Figure 16.1), then the lozenges below the
dents are forced to be vertical (shown as outlined lozenges), and once all of
these forced lozenges are removed, the domain becomes a hexagon.

For the desired $N/2 \times N/2 \times N/2$ hexagon, we choose its location such that
the coordinates of the dents are $-N/2, -N/2 + 1, \ldots, -1$ for the dents on the
right and $-3N/2, \ldots, -N - 1$ for the dents on the left.

Theorem 14.1 from Lecture 14 specializes to the following statement for the
hexagon.[1]

---

[1] Another approach to the explicit inversion of the Kasteleyn matrix for the hexagon is in
Gilmore (2017).

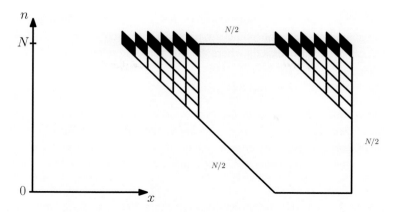

Figure 16.1 A trapezoid/sawtooth domain with the dents chosen such that once the "forced" vertical lozenges are removed, the remainder of the domain is an $N/2 \times N/2 \times N/2$ hexagon. Note that although the diagonal sides appear longer by a factor of $\sqrt{2}$, length should be measured in terms of the sides of our sheared lozenges.

Theorem 16.1 (Petrov, 2014a)   *For even N, consider the $N/2 \times N/2 \times N/2$ hexagon. The inverse Kasteleyn matrix $K^{-1}(\llcorner(y,m); \urcorner(x,n))$ is given by*

$$(-1)^{y-x+m-n+1}\left[\mathbb{1}_{m<n}\mathbb{1}_{y\leq x}\frac{(x-y+1)_{n-m-1}}{(n-m-1)!} \right. \tag{16.1a}$$

$$\left. - \frac{(N-n)!}{(N-m-1)!}\oint_{C_z}\oint_{C_w}\frac{(z-y+1)_{N-m-1}}{(w-x)_{N-n+1}}\frac{(w+1)_{N/2}(w+N+1)_{N/2}}{(z+1)_{N/2}(z+N+1)_{N/2}}\frac{dz\,dw}{w-z}\right], \tag{16.1b}$$

where the contour C_z is a loop around $\{y, y+1, \ldots, -1\}$ (and enclosing no other integers), and the contour C_w is a large loop around C_z and all poles of the integrand.

Because we are sending N to ∞ and studying the local behavior near a point $(N\tilde{x}, N\tilde{n})$ in the bulk, we should work in coordinates that reflect this. Namely, we consider the asymptotic of the inverse Kasteleyn matrix evaluated in two points (x, n) and (y, m) given by

$$x := N\tilde{x} + \Delta x, \tag{16.2}$$

$$y := N\tilde{x} + \Delta y, \tag{16.3}$$

$$n := N\tilde{n} + \Delta n, \tag{16.4}$$

$$m := N\tilde{n} + \Delta m. \tag{16.5}$$

The fact that x and y are both close to $N\tilde{x}$, and similarly, m, n are close to $N\tilde{n}$, guarantees that the points (x, n) and (y, m) stay close to $(N\tilde{x}, N\tilde{n})$. $\Delta x, \Delta y, \Delta n, \Delta m$ control exactly where the two points sit relative to $(N\tilde{x}, N\tilde{n})$ on a local scale.

Here is our aim for this lecture:

Theorem 16.2   *For $(\tilde{x}, \tilde{n})$ inside the liquid region (i.e., inside the ellipse inscribed into the hexagon), the correlation functions of uniformly random tilings of the hexagon converge to those of the ergodic Gibbs translation-invariant measure (see Lecture 13) associated to the slope of the limit shape at $(\tilde{x}, \tilde{n})$.*

Remark 16.3   Because the probabilities of the cylindrical events for random tilings can be expressed as finite linear combinations of the correlation functions, the convergence of the latter implies weak convergence of measures on tilings.

Proof of Theorem 16.2   Theorem 3.5 expresses the correlation functions of tilings through the inverse Kasteleyn matrix that is computed in Theorem 16.1. We would like to apply the steepest-descent method to the double contour integral in (16.1b). For that, we want to write the integrand of (16.1b) as $\exp(N(G(z) - G(w))) \cdot O(1)$ for some function G, where $(z, w) = (N\tilde{z}, N\tilde{w})$. We set

$$G(z) = \frac{1}{N} \sum_{i=1}^{N(1-\tilde{n})} \ln(z - N\tilde{x} + i) - \frac{1}{N} \sum_{i=1}^{N/2} \ln(z+i) - \frac{1}{N} \sum_{i=1}^{N/2} \ln(z + N + i) \quad (16.6)$$

and note that

$$\frac{1}{N} \sum_{i=1}^{N(1-\tilde{n})} \left(\ln(z - N\tilde{x} + i) - \ln N \right) = \frac{1}{N} \sum_{i=1}^{N(1-\tilde{n})} \ln \left(\tilde{z} - \tilde{x} + \frac{i}{N} \right) \quad (16.7)$$

is a Riemann sum for $\int_0^{1-\tilde{n}} \ln(\tilde{z} - \tilde{x} + u)du$. Hence, by subtracting one $\ln N$ term and adding two more to (16.6) to normalize the three sums as in (16.7), we obtain three Riemann sums corresponding to the three integral summands in

$$\tilde{G}(\tilde{z}) := \int_0^{1-\tilde{n}} \ln(\tilde{z} - \tilde{x} + u)du - \int_0^{1/2} \ln(\tilde{z} + u)du - \int_0^{1/2} \ln(\tilde{z} + 1 + u)du.$$

It follows that

$$G(z) = \tilde{G}(\tilde{z}) + o(1) - \tilde{n} \ln N,$$

and the $\ln N$ term will cancel in $G(z) - G(w)$, so we may ignore it.

For steepest-descent analysis, we need the critical points of $\tilde{G}$. In order to compute the derivative $\frac{\partial \tilde{G}}{\partial \tilde{z}}$, we differentiate under the integral sign, then take the definite integral with respect to u which gives us back the logarithm, yielding

$$\frac{\partial \tilde{G}}{\partial \tilde{z}} = \ln(\tilde{z} - \tilde{x} + (1 - \tilde{n})) - \ln(\tilde{z} - \tilde{x}) - \ln(\tilde{z} + 1/2)$$
$$+ \ln(\tilde{z}) - \ln(\tilde{z} + 3/2) + \ln(\tilde{z} + 1).$$

Exponentiating, we get a quadratic equation in $\tilde{z}$ for the critical point:

$$\frac{\partial \tilde{G}}{\partial \tilde{z}} = 0 \iff (\tilde{z} - \tilde{x} + (1 - \tilde{n}))(\tilde{z})(\tilde{z} + 1) = (\tilde{z} - \tilde{x})(\tilde{z} + 1/2)(\tilde{z} + 3/2)$$

$$\iff \tilde{n}\tilde{z}^2 - (\tilde{x} + \frac{1}{4} - \tilde{n})\tilde{z} - \frac{3}{4}\tilde{x} = 0.$$

This equation has a pair of complex conjugate roots if and only if the discriminant is negative:

$$D = (\tilde{x} + 1/4 - \tilde{n})^2 + 3\tilde{n}\tilde{x} < 0.$$

Remark 16.4   The region on the $\tilde{x}, \tilde{n}$ coordinate plane given by $(\tilde{x} + 1/4 - \tilde{n})^2 + 3\tilde{n}\tilde{x} < 0$ is exactly the interior of the inscribed ellipse of the hexagon of side lengths $1/2$, which is the scaled version of Figure 16.1. Clearly, the equation corresponds to a ellipse, and we can check that $(\tilde{x} + 1/4 - \tilde{n})^2 + 3\tilde{n}\tilde{x} = 0$ has a single solution $(-3/4, 1)$ when $\tilde{n} = 1$ and a single solution $(-1/4, 0)$ when $\tilde{n} = 0$, corresponding to the tangency points at the top and bottom of the hexagon. The tangency points to other sides of the hexagon can be found similarly. Because an ellipse is determined by five points, these tangency points fix the ellipse uniquely.

We assume that $D < 0$ and consider two conjugate critical points, z_c and $\bar{z}_c$; this corresponds to $(\tilde{x}, \tilde{n})$ lying in the liquid region.

Next, we wish to find the steepest-descent contours for z and w. For that, we consider level lines of real and imaginary parts ($\Re\tilde{G}(\tilde{z}) = \text{const}$ and $\Im\tilde{G}(\tilde{z}) = \text{const}$) passing through $z_c, \bar{z}_c$. The latter will be the steepest-descent contours, whereas the former will help us to understand their geometry. A schematic drawing of the configuration of the level lines is shown in Figure 16.2. The level lines of the real part, $\Re\tilde{G}(\tilde{z}) = \text{const}$, partition the Riemann sphere into regions where $\Re\tilde{G}(\tilde{z}) - \Re\tilde{G}(z_c)$ is < 0 and > 0, represented by $+$ and $-$ in the figure. Because $\Re\tilde{G}(\tilde{z})$ goes to $-\infty$ as $z \to \infty$, the "outer" region is a "$-$" region, and this determines all the others because crossing a contour changes the sign.

There are two contours passing through z_c and $\bar{z}_c$ along which $\tilde{G}(\tilde{z})$ has a constant imaginary part; one lies in the region of the negative real part, the other

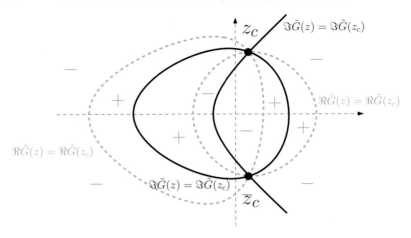

Figure 16.2 Level lines of $\Re\tilde{G}(z)$ and $\Im\tilde{G}(z)$ passing through z_c. The w-contour in (16.1b) is to be deformed to the closed loop of the constant imaginary part in "+" region, and the z-contour is to be deformed to the curve of the constant imaginary part extending to infinity in the "−" region.

in the region of the positive real part. However, the geometry of these contours is not obvious: the one that lies partially in the outer "−" region may either intersect with the positive real axis, intersect with the negative real axis, or just go off to infinity. In order to distinguish between the three possible scenarios, we use the explicit formula for $\tilde{G}(\tilde{z})$. First, note that for complex a, which is very close to the real axis, the value of $\frac{1}{\pi}\Im\ln(z-a)$ is very close to 0 for $z > a$ and close to 1 for $z < a$, with some smoothed-out step-function-like behavior near a. Hence, each of the three $\int\ln$ terms in $\tilde{G}(\tilde{z})$ causes the slope of the function $\frac{1}{\pi}\tilde{G}(x+\mathbf{i}0)$ to change at various points: $\frac{1}{\pi}\int_0^{1-\tilde{n}}\ln(\tilde{z}-\tilde{x}+u)du$ gives the slope a -1 contribution on the segment $[\tilde{x}+\tilde{n}-1,\tilde{x}]$, $-\frac{1}{\pi}\int_0^{1/2}\ln(\tilde{z}+u)du$ gives the slope a 1 contribution on the segment $[-\frac{1}{2},0]$, and $-\frac{1}{\pi}\int_0^{1/2}\ln(\tilde{z}+1+u)du$ gives the slope a 1 contribution on the segment $[-\frac{3}{2},-\frac{1}{2}]$. Adding the three terms, we arrive at the graph of $\frac{1}{\pi}\tilde{G}(x+\mathbf{i}0)$ schematically shown in Figure 16.3.

We are seeking the possible points of intersection of the curve $\Im\tilde{G}(\tilde{z}) = \Im\tilde{G}(z_c)$ with the real axis. They correspond to the intersections of the graph in Figure 16.3 with a horizontal line. Note that the curve $\Im\tilde{G}(\tilde{z}) = \Im\tilde{G}(z_c)$ cannot intersect the real axis in a point inside a horizontal segment of the graph of Figure 16.3 because that would have to be another critical point other than z_c, and we know there are no other critical points. Hence, there are either 1, 3, or 0 points where $\Im\tilde{G}(\tilde{z}) = \Im\tilde{G}(z_c)$ can intersect the real axis.

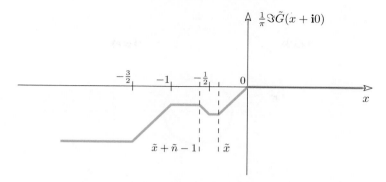

Figure 16.3 The graph of $\Im\tilde{G}(x + i0)$.

Recall that there are four different curves $\Im\tilde{G}(\bar{z}) = \Im\tilde{G}(z_c)$ starting at z_c. If less than three of them reach the real axis, then at least two should escape to infinity. However, this would contradict the configuration of Figure 16.2 because the outer "–" region for the real part can have only one $\Im\tilde{G}(\bar{z}) = \Im\tilde{G}(z_c)$ curve inside (in the upper half-plane). Hence, there are precisely three points of intersection of $\Im\tilde{G}(\bar{z}) = \Im\tilde{G}(z_c)$ with the real axis (each one inside the corresponding slope ±1), and the configuration should necessarily resemble Figure 16.2.

Because we want the integral of $\exp(N(G(z) - G(w)))$ to go to 0, the constant imaginary part contour in the region of the positive real part should be the w contour C_w, and the contour in the negative real part region should be the z contour C_z. Recall that in Theorem 16.1, the two contours look as shown in Figure 16.4 (i.e., one is inside another).

Hence, deforming the integration contour in (16.1b) to be as in Figure 16.2 requires passing C_z through C_w, which picks up the residue at $w = z$ given by

$$\frac{(-1)^{y-x+m-n}}{2\pi\mathbf{i}}\frac{(N-n)!}{(N-m-1)!}\oint_{Nz_c}^{Nz_c}\frac{(z-y+1)_{N-m-1}}{(z-x)_{N-n+1}}dz, \tag{16.8}$$

with the integration contour going to the right from the singularities of the integrand. As we discussed in the previous lecture, the double contour integral on the deformed contours of Figure 16.2 vanishes asymptotically. Hence, the leading contribution of (16.1b) is given by (16.8). Note that in the contour deformation, we should also check that we do not cross any z-poles arising from the denominator $(z + 1)_{N/2}(z + N + 1)_{N/2}$ or w-poles arising from the poles from the denominator $(w - x)_{N-n+1}$; this check can be done by locating

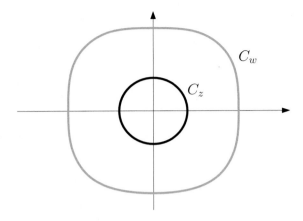

Figure 16.4 In (16.1b) the z-contour is inside the w-contour.

the intersection points of the desired steepest-descent contours with the real line using Figures 16.2 and 16.3.

Sending $N \to \infty$ and substituting (16.2), (16.3), (16.4), (16.5), the integral (16.8) becomes

$$\frac{(-1)^{\Delta x + \Delta n - \Delta y - \Delta m}}{2\pi \mathbf{i}}(1 - \tilde{n})^{\Delta m - \Delta n + 1}$$
$$\times \oint_{\bar{z}_c}^{z_c} (\tilde{z} - \tilde{x})^{\Delta y - \Delta x - 1}(\tilde{z} + \tilde{x} + 1 - \tilde{n})^{\Delta x + \Delta n - \Delta y - \Delta m - 1} d\tilde{z}, \quad (16.9)$$

and the contour again passes to the right from the singularities. We would like to change the variables to simplify (16.9). Set

$$v = \frac{\tilde{z} - \tilde{x}}{\tilde{z} - \tilde{x} + 1 - \tilde{n}}, \tag{16.10}$$

so that

$$1 - v = \frac{1 - \tilde{n}}{\tilde{z} - \tilde{x} + 1 - \tilde{n}}, \tag{16.11}$$

and let v_c be the result when z_c is plugged into the (16.11). Then, changing variables to v in (16.9) yields

$$\frac{(-1)^{\Delta x - \Delta y + \Delta n - \Delta m}}{2\pi \mathbf{i}} \oint_{\bar{v}_c}^{v_c} v^{\Delta y - \Delta x - 1}(1 - v)^{\Delta m - \Delta n} dv, \tag{16.12}$$

where the contour is one for which the intersection with the real axis lies in $(0, 1)$.

We must still deal with the indicator term in (16.1a). Lemma 14.4 says that this term can be identified with the full residue of the double contour integral at $z = w$. One can check that this remains true as $N \to \infty$. Therefore, if we add (16.12) to the indicator function in (16.1a), the result is the *same* integral as in (16.12), except that the integration contour intersects the real axis on $(0, 1)$ for $\Delta m \geq \Delta n$ and intersects on $(-\infty, 0)$ for $\Delta m < \Delta n$, which is when the indicator function is nonzero. This gives us the limiting value of K^{-1}.

Remark 16.5   The limiting kernel that we just found was first introduced in Okounkov and Reshetikhin (2003). It is called the "incomplete beta kernel" (because the integrand is the same as in the classical beta integral) or "extended sine kernel." Here is the reason for the latter terminology.

If we set $\Delta m = \Delta n$ in (16.12), then it becomes

$$\frac{(-1)^{\Delta x - \Delta y}}{2\pi i} \oint_{\bar{v}_c}^{v_c} v^{\Delta y - \Delta x - 1} dv. \tag{16.13}$$

Because the inverse Kasteleyn matrix is encoding probabilities of observing lozenges of different types, for any set of k points with the same n-coordinate and x-coordinates $x_1, \ldots, x_n$, the probability of observing a vertical lozenge (i.e., of the left-most type in the Panel (c) of Figure 14.2) at each of the k points is $\det(K(x_i - x_j))_{1 \leq i,j \leq k}$, where

$$K(x) = \frac{(-1)^x}{2\pi i} \int_{\bar{v}_c}^{v_c} v^{x-1} dv = (-1)^x \frac{(v_c)^x - \bar{v}_c^{\,x}}{2\pi i x} = \frac{(-|v_c|)^x}{\pi x} \sin(x \, \mathrm{Arg}(v_c)). \tag{16.14}$$

The $(-|v_c|)^x$ factor cancels in $\det(K(x_i - x_j))_{1 \leq i,j \leq k}$, and therefore, we can omit it. The conclusion is that the correlation functions of vertical lozenges along a horizontal slice are given by the minors of the matrix

$$K(x) = \begin{cases} \frac{\sin p_1 \pi x}{\pi x}, & x \neq 0, \\ p_1, & x = 0, \end{cases}$$

where p_1 is the asymptotic density of the vertical lozenges. We recognize the discrete sine-kernel, which already appeared before in Corollary 13.16. Hence, the kernel of (16.12) yields the discrete sine kernel when restricted to a line, which is the origin of the name "extended sine kernel."

Exercise 16.6   Show that the measure on lozenge tilings of the plane with correlation functions given by the minors of the extended sine kernel (16.12) (with the choice of integration contour described in the paragraph after the formula) is the same as one coming from the ergodic Gibbs translation-invariant (EGTI) measure of Lecture 13 with slope corresponding to v_c. Note that our ways of drawing lozenges differ between this lecture and Lecture 13 because we used an affine-transformed coordinate system starting from Figure 14.2 in order to match the notations of Petrov (2014a).

This check is a more general version of the argument in Remark 16.5.

Coming back to the proof of Theorem 16.2, at this point, we have established the local convergence of the correlation functions to those of an EGTI measure of Lecture 13. It remains to identify the slope of the local measure with the one of the limit shape for the height function (found through the variational problem and the Burgers equation) at $(\tilde{x}, \tilde{y})$. Indeed, the local slope gives the average proportion of lozenges of each of the three types locally around $(\tilde{x}, \tilde{y})$. Because the asymptotics are uniform over all points $(\tilde{x}, \tilde{y})$, they yield asymptotics for the expected increments of the limiting height function or, equivalently, for the integral of the slope along macroscopic domains. By concentration of the height function, expected increments are asymptotically the same as limit-shape increments. Hence, the integrals of the local slope and of the limit shape slope along macroscopic domains are the same. Therefore, these two kinds of slopes have to coincide.   □

Exercise 16.7 Show that outside the inscribed ellipse, there are six regions (adjacent to six vertices of the hexagon) where one observes only one type of lozenge. Find out which type of lozenge appears in each region, thus explaining Figure 1.5.

17

Lecture 17: Bulk Local Limits Near Straight Boundaries

In the previous lecture we studied the asymptotic behavior of correlation functions for tilings of hexagons. It turned out that in the bulk the correlation functions converge to those of ergodic Gibbs translation-invariant (EGTI) measures. The proof consisted of the steepest descent method applied to the inverse of the Kasteleyn matrix.

We would like to extend this approach to more general regions. To describe the situations where our approach still works, we need the following definition. Recall that *a trapezoid* is a quadrilateral drawn on the triangular grid (with one pair of parallel sides and one pair of nonparallel sides), with the longer base being "dashed" as in Figure 17.1.

Definition 17.1 Let R be a tileable region on the triangular grid. A trapezoid T *is covering a part of R* if

1. $R\backslash T$ touches the boundary of T only along the dashed side;
2. $T\backslash R$ consists of big disjoint triangles along the dashed part of the boundary; that is, $T\backslash R$ can be obtained as the region of T frozen by sequences of consecutive "teeth" along the dashed side.

This definition ensures the following property: every tiling of a region R gives a tiling of a trapezoid T covering R. Note that here, by a tiling of R, we mean a strict tiling, forbidding lozenges to stick out of the region R, whereas a tiling of trapezoid T is allowed to have lozenges sticking out of T (called "teeth") but only along the dashed side (actually, any tiling of T will have exactly N teeth, where N is the height of T); see Figure 17.1.

Theorem 17.2 *Take a sequence of tileable domains $L \cdot R$ with $L \to \infty$ and a point $(x, y) \in R$ such that:*

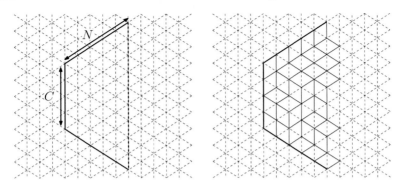

Figure 17.1 A trapezoid with a tiling

1. Proportions $p_{\diamond}, p_{\triangle}, p_{\square}$ encoding the gradient to the limit shape of uniformly random tilings of $L \cdot R$ are continuous and $\neq 0$ in a neighborhood of (x, y).

2. (x, y) is covered by a trapezoid.

Then the correlation functions near (Lx, Ly) converge to those of the EGTI measure of slope $(p_{\diamond}, p_{\triangle}, p_{\square})$.

Remark 17.3   The continuity assumption for $p_{\diamond}, p_{\triangle}, p_{\square}$ ensures that their value at a point is well defined. Also, recall that for polygons, limit shapes are algebraic; hence, the proportions are continuous in the liquid region.

Remark 17.4   For some regions (including a hexagon with a hole), the second condition holds everywhere inside (i.e., the region R can be completely covered by trapezoids). But we cannot completely cover a general region (even a polygon) by trapezoids; see Figure 17.2.

Remark 17.5   The theorem remains valid without the second condition, but the argument becomes much more elaborate (although the steepest-descent analysis of double contour integrals still remains a key ingredient); see Aggarwal (2019).

We will derive Theorem 17.2 as a corollary of the following statement:

Theorem 17.6   *Take a sequence of trapezoids T_N of height N and with teeth $t_1 < \cdots < t_N$. Let*

$$\mu_N = \frac{1}{N} \sum_{i=1}^{N} \delta_{t_i/N}$$

be a scaled empirical measure encoding $\{t_1, \ldots, t_N\}$. Assume the following:

1. The support $\mathrm{Supp}[\mu_n] \subset [-C; C]$ for a constant $C > 0$ independent of N.

2. $\mu_N \xrightarrow[weak]{} \mu$ as $N \to \infty$ for some probability measure μ.

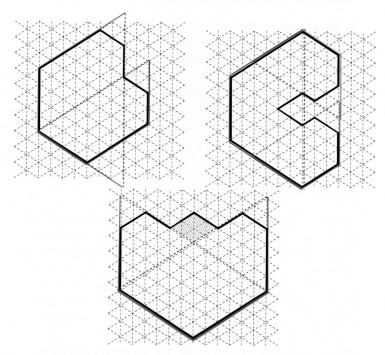

Figure 17.2 Examples of various domains covered by trapezoids. The bottom region cannot be completely covered

Then: 1. Uniformly random tilings of T_N have a limit shape as $N \to \infty$, depending only on μ.

2. For (x, y) such that $p_\diamond, p_\triangle, p_\square$ are continuous and nonzero in a neighborhood of (x, y), correlation functions near (Nx, Ny) converge to those of the EGTI measure with slope $p_\diamond, p_\triangle, p_\square$.

Remark 17.7   In Theorem 17.6, one can take *random* empirical measures μ_N, converging in probability to a deterministic μ. This is exactly the form of Theorem 17.6 used in the proof of Theorem 17.2.

Proof of Theorem 17.2 Using Theorem 17.6   Let (x, y) be a point of R satisfying conditions of the theorem, and let T be a trapezoid covering (x, y). As noted before, tilings of $L \cdot R$ give tilings of $T_{N(L)}$ (for some $N(L)$ tending to infinity as $L \to \infty$) with random teeth $t_1 < \cdots < t_{N(L)}$. Moreover, by the Gibbs property, the restriction of a uniformly random tiling on $L \cdot R$ is a uniformly random tiling on $T_{N(L)}$.

Note that the locations of teeth $t_1 < \cdots < t_{N(L)}$ determine a height function h_L along the dashed side of T_N up to a constant, and $\mu_{N(L)}$ is the distribution of $\frac{h_L(Lx,Ly)}{L}$. Because uniformly random tilings of R converge to a deterministic limit shape, random functions $\frac{h_L(Lx,Ly)}{L}$ converge to a deterministic function $\hat{h}$, and hence the corresponding measures $\mu_{N(L)}$ converge to a deterministic measure μ. Also, note that the length of the dashed side of $T_{N(L)}$ grows linearly as $L \to \infty$; hence, the support of measure $\mu_{N(L)}$ is inside $[-C; C]$ for some $C > 0$.

Thus, both assumptions of Theorem 17.6 hold, and hence the correlation functions of uniformly random tilings near (Lx, Ly) converge to those of EGTI with the same slope. $\square$

Proof of Theorem 17.6   For the first part, note that the convergence of measure μ_n implies the convergence of the height functions on the dashed side of the boundary; hence, the height functions on the boundaries of T_N converge, and we have a limit shape.

The proof of the second part is similar to the hexagon case (which is a particular case where all teeth are concentrated near the end-points of the dashed side): we study the asymptotic behavior of the correlation kernel $K^{-1}((y,m),(x,n))$ using the steepest-descent method. Here, we cover only the main differences from the hexagon case; for the full detailed exposition, see Gorin (2017).

Recall that the correlation kernel is given by the inverse Kasteleyn matrix:

$$(-1)^{y-x+m-n} K^{-1}((y,m),(x,n)) = -\mathbb{1}_{m<n}\mathbb{1}_{y\leq x}\frac{(x-y+1)_{n-m-1}}{(n-m-1)!}$$

$$+ \frac{(N-n)!}{(N-m-1)!}\oint_{C(y,\dots,t_1-1)} dz \oint_{C(\infty)} dw \frac{1}{w-z}\frac{(z-y+1)_{N-m-1}}{(w-x)_{N-n+1}}\prod_{i=1}^{N}\frac{w-t_i}{z-t_i}.$$

The behavior of the indicator part is the same as in the hexagon case, so we will focus on the double integral. Let $\mathcal{T}$ denote the set $\{t_1,\dots,t_N\}$ and $\overline{\mathcal{T}} = \mathbb{Z}\backslash\mathcal{T}$. Then the integrand has the form

$$\frac{1}{(w-x)(w-x+N-n)}\frac{\exp(G_1(z)-G_2(w))}{(w-z)},$$

where

$$G_1(z) = -\sum_{t\in\mathcal{T}}\ln(z-t) + \sum_{i=y-N+m}^{y-1}\ln(z-i)$$

$$= -\sum_{d\in\mathcal{T}\backslash[y-N+m;y-1]}\ln(z-d) + \sum_{d'\in[y-N+m;y-1]\cap\overline{\mathcal{T}}}\ln(z-d'),$$

and $G_2(w)$ is the same as $G_1(z)$ with y, m replaced by x, n.

Recall that in order to use the steepest descent method, we want to find suitable contours such that the exponent will be negative and the whole expression will vanish as $N \to \infty$. To do it, we have taken critical points of G_i as $N \to \infty$ and constructed contours $\Im G_i = $ const passing through them. In the hexagon case, there were only two critical points, but now there can be more, making the construction of the contours seem more complicated. Fortunately, the next two lemmas show that, actually, there are still no more than two nonreal critical points.

Lemma 17.8 *Equation $G_1'(z) = 0$ has either zero or two nonreal roots. Moreover, in the latter case the roots are complex conjugate.*

Proof Rewrite $G_1'(z) = 0$ as

$$-\sum_d \frac{1}{z-d} + \sum_{d'} \frac{1}{z-d'} = 0. \tag{17.1}$$

Let M denote the number of terms in (17.1). Multiplying (17.1) by a common denominator, we get a degree $M-1$ polynomial with real coefficients. Hence, all roots of $G_1'(z) = 0$ are either real or form pairs of complex conjugate roots. Thus, it is enough to show that there always are at least $M-3$ real roots.

Note that the restriction of $G_1'(z)$ to the real line is a real function with M simple poles. Moreover, all poles outside of $[y-N+m; y-1]$ have negative residues, whereas poles inside have positive residue. Then consecutive poles d_i, d_{i+1} such that

$$y - N + m \notin (d_i, d_{i+1}], \quad y - 1 \notin [d_i, d_{i+1})$$

have residues of the same sign. Because there are $M-1$ pairs of consecutive poles and only two pairs can violate the condition on $y-1$ and $y-N+m$, there are $M-3$ pairs of consecutive poles with residues of the same sign. But any continuous real function has a zero between simple poles with the residues of the same sign. Hence, $G_1'(z)$ has at least $M-3$ real roots. $\square$

For $N \to \infty$ set

$$x = N\hat{x} + \Delta x, \quad m = N\hat{n} + \Delta m,$$

$$y = N\hat{x} + \Delta y, \quad n = N\hat{n} + \Delta n.$$

Let

$$\hat{G}(\hat{z}) = \int_{\hat{x}+\hat{n}-1}^{\hat{x}} \ln(\hat{z} - t)dt - \int_{\mathbb{R}} \ln(\hat{z} - t)\mu(t)dt.$$

Lemma 17.9 *$\hat{G}(\hat{z})$ has either zero or two complex conjugate roots.*

Proof Note that

$$G_1'(N\hat{z}) = -\sum_{t \in \mathcal{T}} \frac{1}{N(\hat{z} - t/N)} + \sum_{i=y-N+m}^{y-1} \frac{1}{N(\hat{z} - i/N)}.$$

Both sums are Riemann sums of integrals in

$$\hat{G}'(\hat{z}) = -\int_{\mathbb{R}} \frac{1}{\hat{z} - t} \mu(t) dt + \int_{\hat{x}+\hat{n}-1}^{\hat{x}} \frac{1}{\hat{z} - t} dt.$$

Hence, $G_1'(N\hat{z}) \rightrightarrows \hat{G}'(\hat{z})$ uniformly on compact subsets of $\mathbb{C} \setminus [-D, D]$, where D is a large enough real number. Then, by Hurwitz's theorem, the zeros of $G_1'(N\hat{z})$ converge to zeros of $\hat{G}'(\hat{z})$; hence, $\hat{G}$ has no more than two complex conjugate critical points.   □

Call a pair $(\hat{x}, \hat{n})$ good if there are two complex conjugate critical points of $\hat{G}(\hat{z})$, and let $\tau(\hat{x}, \hat{n})$ denote the critical point in the upper half-plane. Then $\frac{1}{N}G_i(N\hat{z})$ converges to $\hat{G}(\hat{z})$ and as $N \to \infty$, we have

$$\exp(G_1(N\hat{z}) - G_1(N\hat{w})) \approx \exp(N(\hat{G}(\hat{z}) - \hat{G}(\hat{w}))).$$

Now we are in the same situation as in the previous lecture, and repeating the steepest-descent argument, we get (after a change of variables)

$$(-1)^{y-x+m-n} K^{-1}((y,m),(x,n)) \longrightarrow \frac{1}{2\pi\mathbf{i}} \int_{\bar{\xi}}^{\xi} v^{\Delta y - \Delta x - 1} (1-v)^{\Delta m - \Delta n} dv,$$

where $\xi = \frac{\tau - \hat{x}}{\tau - \hat{x} + 1 - \hat{n}}$, and the contour between ξ and $\bar{\xi}$ intersects $\mathbb{R}$ in $[0, 1]$ if $\Delta m \geq \Delta n$ or in $(-\infty; 0)$ otherwise. As noted in the previous lecture, this is exactly the correlation kernel of the EGTI measure. Moreover, at the end of the previous lecture, we proved that the slope of this EGTI measure must match the slope of the limit shape.

To finish the proof, we must show that the "good" points (i.e., the points such that $\tau(\hat{x}, \hat{n})$ is well defined) are exactly the points such that the slope is nonextreme. Recall that the slope $(p_\diamond, p_\square, p_\searrow)$ can be directly recovered from the correlation kernel by looking at the first correlation function; see Corollary 13.16 and Remark 13.17. The three components of the slope (which we intentionally do not write as $\diamond, \square, \searrow$ here in order to avoid the confusion between the two coordinate systems in Lecture 13 and Lectures 14 and 16) are

$$p_1(\hat{x}, \hat{n}) = \lim_{N \to \infty} K^{-1}((N\hat{x}, N\hat{n}), (N\hat{x}, N\hat{n})) = \frac{1}{2\pi\mathbf{i}} \int_{\bar{\xi}+}^{\xi} v^{-1} dv$$

$$= \frac{\ln^+(\xi) - \ln^+(\bar{\xi})}{2\pi\mathbf{i}} = \frac{\operatorname{Arg} \xi}{\pi},$$

$$p_2(\hat{x}, \hat{n}) = \lim_{N \to \infty} K^{-1}((N\hat{x}, N\hat{n}), (N\hat{x} - 1, N\hat{n} + 1)) = \frac{1}{2\pi \mathbf{i}} \int_{\bar{\xi}}^{\xi} (1 - v)^{-1} dv$$

$$= \frac{-\ln^+(1 - \xi) + \ln^+(1 - \bar{\xi})}{2\pi \mathbf{i}} = -\frac{\mathrm{Arg}(1 - \xi)}{\pi},$$

$$p_3(\hat{x}, \hat{n}) = \lim_{N \to \infty} K^{-1}((N\hat{x}, N\hat{n}), (N\hat{x}, N\hat{n} + 1)) = -\frac{1}{2\pi \mathbf{i}} \int_{\bar{\xi}}^{\xi} v^{-1}(1 - v)^{-1} dv$$

$$= -\frac{\ln^-(\frac{\xi}{1-\xi}) - \ln^-(\frac{\bar{\xi}}{1-\bar{\xi}})}{2\pi \mathbf{i}} = 1 - \frac{\mathrm{Arg}(\frac{\xi}{1-\xi})}{\pi},$$

where the contours pass through $[0, 1]$ in the first case and through $(-\infty, 0)$ in the second and third cases, $\ln^+(z)$ is a branch of $\ln(z)$ defined on $\mathbb{C}\backslash(-\infty; 0)$, and $\ln^-(z)$ is a branch defined on $\mathbb{C}\backslash(0, \infty)$.

This computation shows that the three components of the slope (multiplied by Π) are precisely the angles of the triangle with vertices $0, 1, \xi$ (and therefore, ξ is the Kenyon–Okounkov coordinate). Hence for a "good" point, ξ is complex, and the slope is nonextreme. For the other direction, see Gorin (2017).   □

Lecture 18: Edge Limits of Tilings of Hexagons

In this lecture we study the edge limits (as opposed to the bulk limits, which we investigated in several previous lectures) of the uniformly random tilings of the hexagon. We start by explaining what we mean by such limits.

Let us look back at Figures 1.3 and 1.5 in Lecture 1. Near the top of the figures, we see a frozen region filled with green $\diamond$ lozenges. As we move down, we observe the boundary of this region formed by paths built out of red and yellow lozenges of two other types. We would like to understand the scaling limit of this collection of paths right at the frozen boundary or, equivalently, at the *edge* of the liquid region.

We do not need to be close to the frozen region in order to identify paths, as we saw in Lecture 2. Two types of lozenges always form *nonintersecting paths*, as in Figure 18.1. In terms of these paths, we are trying to understand the scaling limit of the top-most path, of the second top-most path and so forth.

Under appropriate rescaling, we expect to see some limiting family of continuous nonintersecting paths. But what scale should we choose to see a nontrivial limit? We need to choose two scalings, one in the direction transversal to the frozen boundary (vertical direction in Figure 18.1) and one in the direction tangential to the frozen boundary (horizontal direction in Figure 18.1).

In the first part of the lecture, we give two heuristics for guessing the transversal and tangential rescalings. Later, we sketch a rigorous analysis.

18.1 Heuristic Derivation of Two Scaling Exponents

For the first heuristic, we notice that the density of the two types of lozenges that we're interested in behaves as $C\sqrt{x}$ (for some constant $C > 0$) in a neighborhood of the edge of the liquid region, where x is the distance from the frozen region rescaled by the linear size L of the hexagon. One way to see this $\sqrt{x}$ behavior is

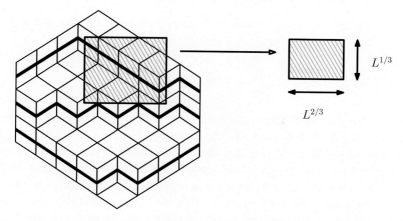

Figure 18.1 A tiling of a hexagon, corresponding nonintersecting paths, and an $L^{2/3} \times L^{1/3}$ window for observing the edge limit.

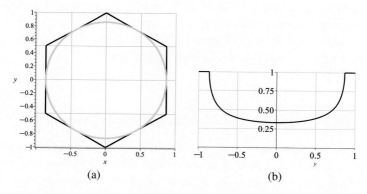

Figure 18.2 Panel (a): Unit hexagon and corresponding frozen boundary. Panel (b): Asymptotic density of horizontal lozenges $\diamond$ along the vertical line $x = 0$. We use formulas from Theorem 1.1 in Cohn et al. (1998).

through the relation of densities to the complex slope. The latter for the hexagon is found as a solution to a quadratic equation; see Lectures 10 and 16. The square root in the formula for the solution of a quadratic equation eventually gives rise to the desired $C\sqrt{x}$ behavior of the densities. Figure 18.2 shows the plot of the complementary density of the lozenges of the third type along one particular vertical section of the hexagon. We remark that the same square-root behavior at the edges also appears in lozenge tilings of more complicated domains.

If we (nonrigorously) assume that the $\sqrt{x}$ behavior of the density extends up to the local scales, then we can deduce one of the transversal edge scalings as a corollary. Indeed, if the kth path is at (vertical) distance x_k from the frozen boundary, then we expect

$$\int_0^{x_k/L} C\sqrt{x}\, dx \approx k/L,$$

implying $x_k = C' \cdot L^{1/3}$. Thus, we expect the kth path to be at a distance of order $L^{1/3}$ from the frozen boundary, and the transversal rescaling should be $L^{1/3}$.

For the second heuristic, note that, locally, paths look like random walks with up-right and down-right steps (these steps correspond to lozenges of types ⬦ and ⬦, respectively). Moreover, there are rare interactions between paths because the spacings are large ($L^{1/3}$ because of the first step). Hence, on the local scale we expect to see some version of a random walk, and therefore, zooming further away, in a proper rescaling, we should locally see some "Brownian-like behavior." For the standard Brownian bridge $B(t)$, $0 \le t \le T$, the variance of $B(\alpha T)$ (for $0 < \alpha < 1$) is $\alpha(1-\alpha)T$, so for large T, one needs a $\sqrt{T}$ transversal rescaling to see nontrivial fluctuations. Because in our case, we already fixed the transversal rescaling to be of order $L^{1/3}$, the horizontal rescaling should be given by a square of that (i.e., $L^{2/3}$), as in Figure 18.1.

18.2 Edge Limit of Random Tilings of Hexagons

For the rigorous part of the lecture, we consider a regular hexagon of side length $N/2$ (for N ranging over even positive integers) and recall the notation and the result of Theorem 16.1 (yet again, we switch from the lozenges of Figure 18.1 to those of Figures 14.2 and 16.1). We know an explicit expression for the probabilities of the kind

Prob (there are lozenges ⬦ at $(x_1, n_1), \ldots, (x_k, n_k)$)
$$= \det\left(K^{-1}((x_i, n_i); (x_j, n_j))\right)_{1 \le i,j \le k}. \quad (18.1)$$

Let us look at the paths formed by two other (not ⬦) types of lozenges. The correlation functions for the points of such paths can be found through the following complementation principle.

Lemma 18.1 *If P is a determinantal point process on a countable set $\mathcal{L}$ (e.g., one can take $\mathcal{L} = \mathbb{Z}$) whose correlation functions are given by the minors of the correlation kernel $C(x, y)$, then the point process of the complementary*

configuration $\overline{P} = \mathcal{L} \setminus P$ is a determinantal point process with the correlation kernel $\overline{C}(x, y) = \delta_{x,y} - C(x, y)$.

Proof Let $k \in \mathbb{N}$ be any positive integer; then

$$\text{Prob}\left(x_1, \ldots, x_k \in \overline{P}\right) = \text{Prob}\,(x_1, \ldots, x_k \notin P)$$

$$= 1 - \sum_{1 \le i \le k} \text{Prob}\,(x_i \in P) + \sum_{\substack{1 \le i,j \le k \\ i \ne j}} \text{Prob}\left(x_i, x_j \in P\right) - \cdots$$

$$= 1 - \sum_{1 \le i \le k} C(x_i, x_i) + \sum_{\substack{1 \le i,j \le k \\ i \ne j}} \det \begin{bmatrix} C(x_i, x_i) & C(x_i, x_j) \\ C(x_j, x_i) & C(x_j, x_j) \end{bmatrix} - \cdots$$

$$= \det \left[\delta_{i,j} - C(x_i, x_j)\right]_{1 \le i,j \le k},$$

where the first equality is obvious, the second is the principle of inclusion–exclusion, the third is because P is a determinantal point process, and the last equality follows from using the multilinearity (and definition) of the determinant. $\square$

From (18.1) and the previous lemma, we have

Prob (non intersecting paths go through $(x_1, n_1), \ldots, (x_k, n_k)$)

$$= \det \left(\delta_{i,j} - K^{-1}((x_i, n_i); (x_j, n_j))\right)_{1 \le i,j \le k}. \quad (18.2)$$

Thus, we would like to compute the limit of $\delta_{(x_1,n_1),(x_2,n_2)} - K^{-1}((x_1, n_1); (x_2, n_2))$ when (x_1, n_1) and (x_2, n_2) are close to the boundary of the frozen region.

Recall the result of Theorem 16.1:

$$K^{-1}((x_1, n_1); (x_2, n_2))$$

$$= (-1)^{x_1 - x_2 + n_1 - n_2 + 1} \mathbf{1}_{\{n_1 < n_2\}} \mathbf{1}_{\{x_1 \le x_2\}} \frac{(x_2 - x_1 + 1)_{n_2 - n_1 - 1}}{(n_2 - n_1 - 1)!}$$

$$+ (-1)^{x_1 - x_2 + n_1 - n_2} \frac{(N - n_2)!}{(N - n_1 + 1)!} \oint_{C(x_1, x_1 + 1, \ldots, t_1 - 1)} \frac{dz}{2\pi \mathbf{i}} \oint_{C(\infty)} \frac{dw}{2\pi \mathbf{i}}$$

$$\times \frac{1}{w - z} \frac{(z - x_1 + 1)_{N - n_1 - 1}}{(w - x_2)_{N - n_2 + 1}} \frac{(w + 1)_{N/2}(w + N + 1)_{N/2}}{(z + 1)_{N/2}(z + N + 1)_{N/2}}. \quad (18.3)$$

We will zoom in near the right part of the frozen boundary, circled in Figure 18.3. From our initial heuristics, we want the points (x_1, n_1) and (x_2, n_2) to be separated by the distance $O(N^{2/3})$ in the tangential direction to the frozen boundary and by $O(N^{1/3})$ in the normal direction.

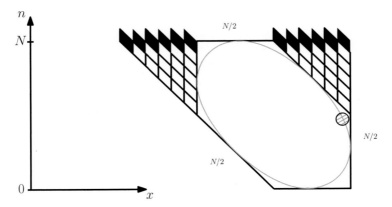

Figure 18.3   The hexagonal domain, the boundary of the frozen region, and the part of the boundary where we zoom in to see the edge limit.

For that, we specify the points via

$$n_1 = \widetilde{n}N + (\Delta n_1)N^{2/3}, \quad n_2 = \widetilde{n}N + (\Delta n_2)N^{2/3},$$
$$x_1 = x(n_1) + (\Delta x_1)N^{1/3}, \quad x_2 = x(n_2) + (\Delta x_2)N^{1/3}. \tag{18.4}$$

At this moment, we need to be careful in specifying $x(n_1)$, $x(n_2)$ – the *curvature* of the frozen boundary is important (because of the different tangential and normal scalings), and we are not allowed to simply use linear approximation.

Taking the logarithm of the z-part of the integrand in (18.3), we define $G_N^{n_1}(z)$ through

$$G_N^{n_1}(z) := \sum_{i=1}^{N-n_1} \ln(z - x + i) - \sum_{i=1}^{N/2} \ln(z + i) - \sum_{i=1}^{N/2} \ln(z + N + i).$$

The equation $\frac{\partial}{\partial z} G_N^{n_1}(z) = 0$ is

$$\frac{\partial}{\partial z} \left(\sum_{i=1}^{N-n_1} \ln(z - x + i) - \sum_{i=1}^{N/2} \ln(z + i) - \sum_{i=1}^{N/2} \ln(z + N + i) \right) = 0.$$

Instead of this equation, it is easier to consider its $N \to \infty$ approximation:

$$\frac{\partial}{\partial z} \left(\int_0^{1 - \frac{n_1}{N}} \ln \left(\frac{z}{N} - \frac{x}{N} + u \right) du \right.$$
$$\left. - \int_0^{1/2} \ln \left(\frac{z}{N} + u \right) du - \int_0^{1/2} \ln \left(\frac{z}{N} + 1 + u \right) du \right) = 0.$$

This equation is quadratic in z, and we need to take $x = x(n_1)$ so that this equation has a double real root z_c. The value of $x(n_2)$ is obtained through the same procedure.

The steepest-descent analysis of the double contour integral is done similarly to the bulk case, except that now two complex conjugate critical points merge into a single double critical point on the real axis. In the neighborhood of the critical point, we have

$$G_N^{n_1}(z) \approx G_N^{n_1}(z_c) + (z - z_c)^3 \alpha, \tag{18.5}$$

for some explicit α.

Previously, in the bulk analysis, the steepest-descent contours of integration passed through two critical points, but now there is only one critical point z_c, so the situation is degenerate. One can obtain the configuration of the contours for the present case by continuously deforming Figure 16.2 into the situation of merging critical points. The z contour should pass through z_c, and the w contour should pass through the similarly defined w_c (which is close to z_c because n_2 is close to n_1) under such angles as to guarantee the negative real part of $G_N^{n_1}(z) - G_N^{n_2}(w)$ after using (18.5) in a neighborhood of the critical point.

For the limit, we zoom in near the critical points, where the contours can be chosen to look like they do in Figure 18.4. As in the bulk case, we also need to make sure that this local picture can be extended to the global steepest-descent contours (such that the real part of the exponent is still maximized near the critical point), but we omit this part of the argument. We reach the following theorem and refer the reader to Section 8 in Petrov (2014a) for more details.

Theorem 18.2 *In the limit regime* (18.4), *which describes the random lozenge tiling near the boundary of the frozen region, we have*

$$\lim_{N \to \infty} N^{1/3} \left(\delta_{(x_1, n_1), (x_2, n_2)} - K^{-1}((x_1, n_1); (x_2, n_2)) \right) = \mathcal{A}((\Delta x_1, \Delta n_1); (\Delta x_2, \Delta n_2)),$$

where

$$\mathcal{A}((\Delta x_1, \Delta n_1); (\Delta x_2, \Delta n_2)) := -\mathbf{1}_{\Delta n_1 < \Delta n_2} (\text{residue of the integral at its pole})$$

$$+ \frac{1}{(2\pi \mathbf{i})^2} \int \int \exp\left(\frac{\widetilde{w}^3 - \widetilde{z}^3}{3}\right) \exp\left(-\widetilde{w} \Delta x_2 \beta + \widetilde{z} \Delta x_1 \beta\right) \frac{\beta \, d\widetilde{w} \, d\widetilde{z}}{\widetilde{w} - \widetilde{z} + \gamma(\Delta n_1 - \Delta n_2)}, \tag{18.6}$$

and where β, γ are explicit (by rescaling $\Delta x_i, \Delta n_j$ properly, these constants can actually be removed from the formula). The contours are as in Figure 18.4.

Remark 18.3 The z and w integration variables were shifted by different amounts z_c and w_c when getting (18.6). Hence $\frac{1}{w-z}$ turned into $\frac{1}{\widetilde{w}-\widetilde{z}+\gamma(\Delta n_1 - \Delta n_2)}$.

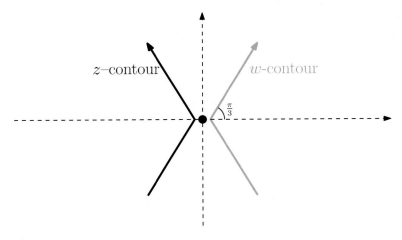

Figure 18.4 Local configuration of the steepest-descent integration contours for the edge limit. The figure is centered at the critical point.

The tips of the contours (where they intersect the real axis) should have sufficient distance between themselves so that the denominator does not vanish on the contours. The factor $\exp\left(-\widetilde{w}\Delta x_2\beta + \widetilde{z}\Delta x_1\beta\right)$ in (18.6) appears because $G_N^{n_1}(z)$ has $x = x(n_1)$ in its definition, but in the integrand of (18.3), we instead have $x = x_1$ (and similarly for $G_N^{n_2}(w)$).

Remark 18.4   The convergence of the correlation kernels implies convergence of the paths in the sense of finite-dimensional distributions. The limiting object is defined through its correlation function given by the minors of (18.6). The definition of the correlation functions is slightly more complicated for the continuous (rather than discrete) state space; see, for example, Johansson (2006), Borodin (2011b), or Section 3 in Borodin and Gorin (2016) for the details in determinantal contexts, and see Daley and Vere-Jones (2003) for the general theory.

Exercise 18.5   Finish the proof of Theorem 18.2 and compute the values of β and γ.

A rescaled version of the kernel $\mathcal{A}$ in (18.6) is known as the "extended Airy kernel" in the literature. Here is its standard form:

$$K_{\text{extended}}^{\text{Airy}}\left((x,t);(y,s)\right) := \begin{cases} \int_0^{\infty} e^{-\lambda(t-s)}\text{Ai}(x+\lambda)\text{Ai}(y+\lambda)d\lambda, & \text{if } t \geq s, \\ -\int_{-\infty}^{0} e^{-\lambda(t-s)}\text{Ai}(x+\lambda)\text{Ai}(y+\lambda)d\lambda, & \text{if } t < s, \end{cases}$$

$$(18.7)$$

where

$$\mathrm{Ai}(z) := \frac{1}{2\pi\mathbf{i}} \int \exp\left(\frac{v^3}{3} - zv\right) dv \tag{18.8}$$

is the Airy function[1]; the contour in the integral is the upward-directed contour that is the union of the lines $\{e^{-\mathbf{i}\pi/3}t : t \ge 0\}$ and $\{e^{\mathbf{i}\pi/3}t : t \ge 0\}$ (it looks like the w-contour in Figure 18.4).

Exercise 18.6 Match $K_{\text{extended}}^{\text{Airy}}((x,t);(y,s))$ with $\mathcal{A}((x,t);(y,s))$ in Theorem 18.2 by substituting (18.8) into (18.7) and explicitly evaluating the λ-integral.

The conclusion is that the ensembles of extreme paths (edge limit) in random tilings of large hexagons converge to the random path ensemble whose correlation functions are minors of the extended Airy kernel.

18.3 The Airy Line Ensemble in Tilings and Beyond

It is believed that Theorem 18.2 is not specific to hexagons.

Conjecture 18.7 *The extended Airy kernel arises in the edge limit of the uniform lozenge tilings of arbitrary polygonal domains as the mesh size of the grid goes to zero.*

The conjecture has been proven in certain cases (for trapezoids) in Petrov (2014a) and Duse and Metcalfe (2018). The first appearance of the extended Airy kernel is in Prähofer and Spohn (2002) in the study of the polynuclear growth (PNG) model. Closer to the tilings context, Ferrari and Spohn (2003) were the first to find the extended Airy kernel in the edge limit of q^{Volume}-weighted plane partitions (we will discuss this model of random plane partitions in Lecture 22), and Johansson (2005) found it in the domino tilings of the Aztec diamonds.

The random two-dimensional ensemble (a random ensemble of paths) arising from the extended Airy kernel is called the "Airy line ensemble." It is a translation-invariant ensemble (in the space direction). In the next lecture, we will learn more about it and tie it to other subjects in probability. As a warm-up, let us mention that the one-point distribution of the "top-most" path is known as the "Tracy–Widom (TW) distribution."[2] This distribution also appears in various other contexts:

---

[1] The Airy function solves the second-order differential equation $\frac{\partial^2}{\partial z^2}\mathrm{Ai}(z) - z\mathrm{Ai}(z) = 0$.

[2] The Wikipedia page https://en.wikipedia.org/wiki/Tracy-Widom_distribution is a good initial source of information.

1. In random matrix theory, the TW distribution is (the limit of) the distribution of the largest eigenvalue of large Hermitian random matrices of growing sizes; see the reviews in Forrester (2010), Anderson et al. (2010), and Akemann et al. (2011). In the next two lectures, we will explain the link between tilings and random matrices in more detail.

2. A closely related applied point of view says that the TW distribution appears as an asymptotic law in statistical hypothesis-testing procedures involving sample covariance matrices, sample canonical correlations between various types of data, and multivariate analysis of variance (MANOVA). Note that in order to see our TW distribution in such a context, one needs to deal with *complex data*, which might be less common. The more widespread setting of real data leads to the appearance of the closely related $\beta = 1$ Tracy–Widom distribution (TW_1); see, for example, Johnstone (2008) and Bykhovskaya and Gorin (2020) for examples of such results.

3. In interacting particle systems, the TW distribution arises as the limiting particle current for various systems, including the totally asymmetric simple exclusion process (TASEP). We do not have time to discuss the direct links between tilings and interacting particle systems in these lectures, and we instead only refer the reader to the reviews in (Johansson, 2006), (Borodin and Gorin, 2016), (Borodin and Petrov, 2014), and (Romik, 2015).

4. In directed polymers, the TW distribution governs the asymptotic law of the partition function for directed polymers in random media and the related solution of the Kardar–Parisi–Zhang stochastic partial differential equation; see the reviews in Corwin (2012) and (Quastel and Spohn, 2015)

5. In asymptotic representation theory, the TW distribution is the asymptotic law for the first row of random Young diagrams encoding the decomposition of representations (typically of unitary or symmetric groups) into irreducible components. The most celebrated result of this kind deals with the Plancherel measure for the symmetric groups; see the reviews in Johansson (2006), Borodin and Gorin (2016), Romik (2015), and Baik et al. (2016).

Lecture 19: The Airy Line Ensemble and Other Edge Limits

19.1 Invariant Description of the Airy Line Ensemble

In the previous lecture we found the Airy line ensemble as the edge limit of the uniformly random tilings of hexagons. We described it in terms of its correlation functions, which are minors of the extended Airy kernel of (18.7). On the other hand, in the bulk, we could describe the limiting object without formulas: it was a unique ergodic Gibbs translation-invariant (EGTI) measure of a given slope.

Is there a similar description for the Airy line ensemble?

Conjecturally, the answer is yes, based on the following three facts:
1. The Airy line ensemble is *translation-invariant* in the t (i.e., horizontal) direction – this immediately follows from (18.7).

2. The Airy line ensemble possesses an analogue of the Gibbs property known as the "Brownian Gibbs property." Let $\{\mathcal{L}_i(t)\}_{i\geq 1}$ be an ensemble of random paths such that $\mathcal{L}_1(t) \geq \mathcal{L}_2(t) \geq \cdots$, for all $t \in \mathbb{R}$. It satisfies the Brownian Gibbs property if the following is true: Fix arbitrary $k = 1, 2, \ldots$, two reals $a < b$, k reals $x_1 > \cdots > x_k$, k reals $y_1 > \cdots > y_k$, and a trajectory $f(t)$, $a \leq t \leq b$, such that $f(a) < x_k$ and $f(b) < y_k$. Then the conditional distribution of the first k paths $\mathcal{L}_1, \ldots, \mathcal{L}_k$ conditioned on $\mathcal{L}_i(a) = x_i$, $\mathcal{L}_i(b) = y_i$, for $1 \leq i \leq k$, and on $\mathcal{L}_{k+1}(t) = f(t)$, for $a \leq t \leq b$, is the same as that of k independent Brownian bridges conditioned to have no intersection and to stay above $f(t)$, $a \leq t \leq b$; see Figure 19.1.

Theorem 19.1 (Corwin and Hammond, 2014) *For the Airy line ensemble $\mathcal{A}_i(t)$, $i = 1, 2, \ldots$, with correlation functions given by the minors of (18.7), let $\mathcal{L}_i(t) := \frac{1}{\sqrt{2}}(\mathcal{A}_i(t) - t^2)$, for all $i \geq 1$. Then the random path ensemble $\{\mathcal{L}_i\}_{i\geq 1}$ satisfies the Brownian Gibbs property.*

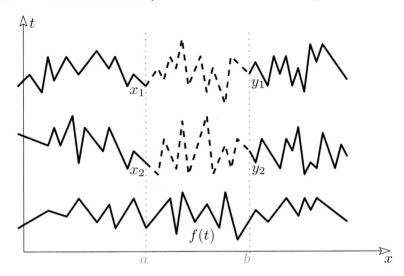

Figure 19.1 The Brownian Gibbs property: Given x_1, x_2, y_1, y_2 and $f(t)$, $a \leq t \leq b$, the distribution of the dashed portion of the paths is that of nonintersecting Brownian bridges staying above $f(t)$.

Roughly speaking, the proof is based on the Gibbs resampling property for the tilings: clearly, the nonintersecting paths coming from uniformly random tilings, as in Figure 18.3, satisfy a similar Gibbs property with Brownian bridges replaced by simple random walks. We know that these paths converge to the Airy line ensemble, and simultaneously, nonintersecting random walks converge to nonintersecting Brownian bridges in the same regime. What remains to be shown (and what is the essence of the Corwin and Hammond (2014) argument) is that the topology of the convergence toward the Airy line ensemble can be made strong enough to guarantee that the Gibbs property survives in the limit. The subtraction of t^2 in the statement comes from the curvature of the frozen boundary near the edge, which was a part of $x(n)$ in (18.4).

3. Corwin and Sun (2014) proved that the sequence $\{\mathcal{A}_i(t) + C\}_{i \geq 1}$ is a translation-invariant, *ergodic* line ensemble for any constant C.

The additive constant C mentioned in the last point plays the role of the slope (s, t) for the corresponding result at the bulk. Taking it into account, the following conjecture was formulated by Okounkov and Sheffield in the 2000s:

Conjecture 19.2 $\{\mathcal{A}_i(t)\}_i$ *and its deterministic shifts (in the vertical direction) are the unique translation-invariant, ergodic line ensembles such that the sequence* $\{\frac{1}{\sqrt{2}}(\mathcal{A}_i(t) - t^2)\}_{i \geq 1}$ *has the Brownian Gibbs property.*

Conceptual conclusion. The Airy line ensemble should appear whenever we deal with a scaling limit of random ensembles of nonintersecting paths at the points where their density is low.

19.2 Local Limits at Special Points of the Frozen Boundary

In the previous lecture we saw the appearance of the Airy line ensemble at a *generic* point where the frozen boundary is smooth. However, as we tile various complicated polygons, the frozen boundary develops various singularities, where the asymptotic behavior changes.

- One type of special point of the boundary arises in any polygon. These are the points where the frozen boundary is tangent to a side of the polygon. For instance, in a hexagon, we have six such points. In this situation, the correct scaling factor in the tangential direction to the frozen boundary is $L^{1/2}$ (where L is the linear size of the domain), and we do not need any rescaling in the normal direction. The limiting object *Gaussian unitary ensemble (GUE)-corners process* connects us to the random matrix theory, and we will discuss it in the next section and continue in the next lecture.

- The frozen boundary can develop cusp singularities. For instance, they are visible in the connected component of the boundary surrounding the hole in our running example of the holey hexagon; see Figure 1.3 in Lecture 1 and Figures 24.1 and 24.5 in Lecture 24. There are two types of these cusps: one type is presented by two vertical cusps in the previously mentioned figures, which separate two types of frozen regions; the limit shape develops a corner at such a cusp. In this situation, the scaling factors are $L^{1/3}$ in the direction of the cusp and finite in the orthogonal direction. The scaling limit is known as a "Cusp–Airy process"; see Okounkov and Reshetikhin (2006) and Duse et al. (2016). We also observe two cusps of another type in the picture: these cusps are each adjacent to a single type of the frozen region; the limit shape is $\mathbb{C}^1$-smooth near these cusps. The fluctuations of the discrete boundaries of such a cusp are of order $L^{1/4}$, and one needs to travel the distance of order $L^{1/2}$ along the cusp to see a nontrivial two-dimensional picture. The limit is known as the (extended) *Pearcey process*; see Okounkov and Reshetikhin (2007) and Borodin and Kuan (2008) for its appearances in tilings.

- Another possible situation is when the frozen boundary (locally near a point) has two connected components that barely touch each other. As in the cusp case, there are two subcases depending on how many types of frozen regions we see in a neighborhood, and we refer the reader to Adler et al. (2018a) and Adler et al. (2018b) and the references therein for exact results. The limiting processes that appear in this situation share the name "Tacnode process."

Together with generic points, the previous list gives six different scaling limits, corresponding to six ways the frozen boundary can appear near a given point. As Section 9 in Astala et al. (2020) suggests, this list corresponds to six types of local behaviors of the frozen boundaries, and as long as we deal with uniformly random tilings of polygons with macroscopic side lengths, no other types of behavior for the frozen boundary should occur.

In all these situations, the processes that appear in the limit are believed to be universal (and their universality extends beyond random lozenge tilings), but such asymptotic behavior has been proven rigorously only for specific situations.

Let us also remark that if we allow nonuniform measures on tilings,[1] then the number of possibilities starts to grow, and we can encounter more and more exotic examples. Just to give one example: Borodin et al. (2010) studied tilings with weight proportional to the product $w(\diamond)$ over all horizontal lozenges in tilings. The possible choices for $w(\cdot)$ included the linear function of the vertical coordinate of the lozenge $\diamond$. Tuning the parameters, one can make this weight vanish near one corner of the hexagon, leading to the frozen boundary developing a node; see Figure 19.2. Again, one would expect to see a new scaling limit near this node.

19.3 From Tilings to Random Matrices

Our next topic is the link between random tilings and *random matrix theory*. Indeed, historically, the distribution $\mathcal{A}_1(0)$ of the "top-most" path of the Airy line ensemble at time $t = 0$, known as the "Tracy–Widom (TW) distribution," first appeared in the random matrix theory. At this point, one might be wondering what tilings have in common with matrices and why the same limiting objects appear in both topics, and we are going to explain that now.

The general philosophy comes from the notion of the *semiclassical limit* in representation theory: *large representations of a Lie group G behave like orbits of the coadjoint action on the corresponding Lie algebra* $\mathfrak{g}$.

---

[1] Alternatively, one can tile more complicated domains than macroscopic polygons (e.g., by allowing various tiny defects in the domain boundaries).

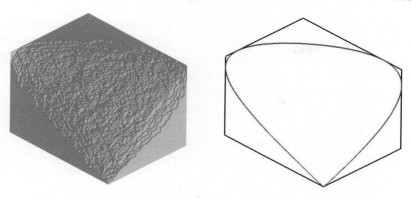

Figure 19.2 Random lozenge tiling of a $70 \times 90 \times 70$ hexagon and corresponding limit shape. The weight of tiling is proportional to the product of the vertical coordinates of all its horizontal lozenges (shown in green).

When $G = U(N)$, the Lie algebra is $\mathfrak{g} = i\text{Herm}(N) \cong \text{Herm}(N)$ (the latter is the space of complex Hermitian matrices). We care about this case because tilings in the hexagonal lattice are linked to the representation theory of $U(N)$. This is illustrated by the following proposition, which claims that the number of tilings of a certain domain can be calculated via Weyl's formula for dimensions of $U(N)$ irreducible representations.

Proposition 19.3 *The number of tilings of the trapezoid with teeth at positions $t_1 < t_2 < \cdots < t_N$, as in Figures 14.1 and 14.2, is*

$$\prod_{1 \leq i < j \leq N} \frac{t_j - t_i}{j - i}$$

and coincides with the dimension of the irreducible representation of $U(N)$ with signature (highest weight) $\lambda_1 \geq \lambda_2 \geq \cdots \geq \lambda_N$, such that $t_{N+1-i} = \lambda_i + N - i$, $i = 1, \ldots, N$.

Proof This is left as an exercise; we only mention three possible approaches:

- Identify the terms in the *combinatorial formula* for the expansion of Schur polynomials $s_\lambda(x_1, x_2, \ldots, x_N)$ into monomials with tilings of the trapezoid. Then, use known formulas for the evaluation of Schur polynomials $s_\lambda(1, 1, \ldots, 1)$. We refer the reader to Chapter I in Macdonald (1995) for extensive information on the Schur polynomials and to Section 2.2 in Borodin and Petrov (2014) for a review of the interplay between tilings and symmetric polynomials. This interplay appears again in Lecture 22 and is explained there in more detail.

- Identify tilings of the trapezoid with basis elements of the irreducible representation V_λ of $U(N)$ associated to λ. Then use Weyl's formula to find the dimension of the irreducible representation of $U(N)$ corresponding to λ. See, for example, Section 2.2 in Borodin and Petrov (2014) for more details on this approach and further references.
- Identify tilings of the trapezoid with families of nonintersecting paths, as in the previous lecture. Then use Theorem 2.5 from Lecture 2 to count the number of such nonintersecting paths via determinants, and use the evaluation formulas from Krattenthaler (1999a) to compute these determinants.   □

Exercise 19.4   Identifying a hexagon with a trapezoid as in Figure 18.3, use Proposition 19.3 to give another proof of the MacMahon formula (1.1) for the total number of tilings of an $A \times B \times C$ hexagon.

Remark 19.5   The enumeration of tilings of trapezoids possessing *axial symmetry* also has a connection to representation theory, this time of orthogonal and symplectic groups. We refer the reader to Borodin and Kuan (2010) and Section 3.2 in Bufetov and Gorin (2015) for some details.

We now explain an instance of the semiclassical limit by making a down-to-earth computation for the random lozenge tilings of a hexagon. For that, let us investigate the law of the N horizontal lozenges at distance N from the left edge in a uniformly random tiling of the $A \times B \times C$ hexagon, as in Figure 19.3. The vertical line at distance N from the left intersects two sides of the hexagon and splits it into two parts. The left part has the geometry of a trapezoid. The right part can also be identified with a trapezoid if we extend it to the continuation of two sides of the trapezoid, as in Panel (b) of Figure 19.3.

Using Proposition 19.3 and denoting the positions of the desired N horizontal lozenges through $t_1, \ldots, t_N$, we compute the following for the case $N < A = B = C$ (the general case can be obtained similarly):

$$\text{Prob}(t_1, \ldots, t_N) = \frac{\#\text{tilings of left trapezoid} \times \#\text{tilings of right trapezoid}}{\#\text{tilings of the hexagon}}$$

$$\propto \prod_{1 \le i < j \le N} (t_j - t_i) \cdot \prod_{1 \le a < b \le 2A - N} (\widetilde{t}_b - \widetilde{t}_a)$$

$$\propto \prod_{1 \le i < j \le N} (t_j - t_i)^2 \cdot \prod_{i=1}^{N} (t_i + 1)_{A-N} (A + N - t_i)_{A-N}.$$

$$(19.1)$$

(Recall the Pochhammer symbol notation: $(x)_n := x(x+1)\cdots(x+n-1) = \Gamma(x+n)/\Gamma(x)$.)

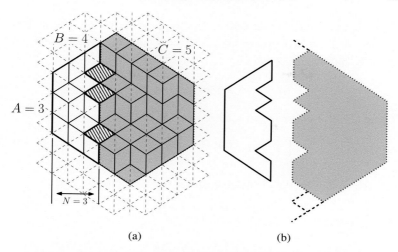

(a)   (b)

Figure 19.3 Three horizontal lozenges at the vertical line at distance 3 of the left edge in a $3 \times 4 \times 5$ hexagon. In order to find the law of these lozenges, one needs to count the tilings of two trapezoids: to the left and to the right from the cut.

Here, $0 \le t_1 < \cdots < t_N \le A + N - 1$ are the positions of the horizontal lozenges at level N. Equivalently, they are the teeth of the left trapezoid. On the other hand, $\widetilde{t}_1 < \widetilde{t}_2 < \cdots < \widetilde{t}_{2A-N}$ are the positions of the teeth of the right trapezoid. That is, $\{\widetilde{t}_1, \ldots, \widetilde{t}_{A-N}\} = \{-1, \ldots, -(A - N)\}$, $\{\widetilde{t}_{A+1}, \ldots, \widetilde{t}_{2A-N}\} = \{A + N, \ldots, 2A - 1\}$ and $\{\widetilde{t}_{A-N+1}, \ldots, \widetilde{t}_A\} = \{t_1, \ldots, t_N\}$.

Exercise 19.6 Generalize the computation (19.1) to arbitrary $A \times B \times C$ hexagons and arbitrary values of $0 \le N \le B + C$.

Exercise 19.7 Find an analogue of (19.1) for q^{Volume}-weighted lozenge tilings of the hexagon.[2]

We would like to study the limit $A \to \infty$ of the distribution (19.1) while keeping N fixed. It is clear that $\mathbb{E}[t_i] \approx A/2$ when A is large compared to N because of the symmetry of the model. Hence, recentering by $\frac{A}{2}$ is necessary. Let us guess the magnitude of the fluctuations by looking at the simplest $N = 1$ case. From (19.1), we have $\text{Prob}(t) \propto (t + 1)_{A-1}(A + 1 - t)_{A-1}$, which is proportional to the product of two binomial coefficients. Recall the De Moivre–Laplace central limit theorem, which says that if $\text{Prob}(t)$ is proportional to a single binomial coefficient, then the order of the fluctuations, as $A \to \infty$, would

---

[2] The computation is possible for more complicated distributions on tilings as well. In particular, Section 10 in Borodin et al. (2010) studies the situation where the weights of lozenges are given by elliptic functions in their coordinates.

be $\sqrt{A}$. It turns out that the same is true in our case of the product of two binomial coefficients. Thus, we expect to see a nontrivial scaling limit in the following regime:

$$t_i = \frac{A}{2} + \sqrt{A}x_i, \quad i = 1, \ldots, N. \tag{19.2}$$

We plug (19.2) into (19.1) and send $A \to \infty$. With (19.2), we have

$$\prod_{1 \le i < j \le N} (t_i - t_j)^2 \propto \prod_{1 \le i < j \le N} (x_i - x_j)^2, \tag{19.3}$$

where the hidden constant is independent of the x_i. Moreover,

$$(t_i + 1)_{A-N}(A + N - t_i)_{A-N} = \frac{(t_i + A - N)!}{t_i!} \frac{(2A - t_i - 1)!}{(A + N - t_i - 1)!}$$

$$= \frac{\left(\frac{3A}{2} - N + \sqrt{A}x_i\right)!}{\left(\frac{A}{2} + \sqrt{A}x_i\right)!} \frac{\left(\frac{3A}{2} - \sqrt{A}x_i - 1\right)!}{\left(\frac{A}{2} + N - \sqrt{A}x_i - 1\right)!}.$$

We can use Stirling's formula $M! = \sqrt{2\pi M}(M/e)^M(1 + o(1))$, $M \to \infty$ in the previous formula. It is safe to ignore N and other finite-order constants for the computation because only growing terms will give a nontrivial contribution. Using the asymptotic expansion $\ln(1 + x) = x - x^2/2 + O(x^3)$, $x \to 0$, we get

$$(t_i + 1)_{A-N}(A + N - t_i)_{A-N}$$

$$\approx \exp\left(\left(\frac{3A}{2} + \sqrt{A}x_i\right)\left(\ln\left(\frac{3A}{2} + \sqrt{A}x_i\right) - 1\right) + \left(\frac{3A}{2} - \sqrt{A}x_i\right)\left(\ln\left(\frac{3A}{2} - \sqrt{A}x_i\right) - 1\right)\right)$$

$$\exp\left(-\left(\frac{A}{2} + \sqrt{A}x_i\right)\left(\ln\left(\frac{A}{2} + \sqrt{A}x_i\right) - 1\right) - \left(\frac{A}{2} - \sqrt{A}x_i\right)\left(\ln\left(\frac{A}{2} - \sqrt{A}x_i\right) - 1\right)\right)$$

$$\approx \exp\left(A\left[\left(\frac{3}{2} + \frac{x_i}{\sqrt{A}}\right)\left(\ln\left(1 + \frac{2x_i}{3\sqrt{A}}\right)\right) + \left(\frac{3}{2} - \frac{x_i}{\sqrt{A}}\right)\ln\left(1 - \frac{2x_i}{3\sqrt{A}}\right)\right]\right)$$

$$\exp\left(A\left[-\left(\frac{1}{2} + \frac{x_i}{\sqrt{A}}\right)\left(\ln\left(1 + \frac{2x_i}{\sqrt{A}}\right)\right) - \left(\frac{1}{2} - \frac{x_i}{\sqrt{A}}\right)\ln\left(1 - \frac{2x_i}{\sqrt{A}}\right)\right]\right)$$

$$\approx \exp\left(-\frac{4}{3}x_i^2\right).$$

As a result, in the regime in (19.2), we have

$$\prod_{i=1}^{N} (t_i + 1)_{A-N}(A + N - t_i)_{A-N} = \prod_{i=1}^{N} \exp\left(-\frac{4}{3}x_i^2\right) \cdot (1 + o(1)). \tag{19.4}$$

From (19.1), (19.3), and (19.4), we get the following proposition, whose proof first appeared in Nordenstam (2009):

Proposition 19.8 *Consider uniformly random tilings of an $A \times A \times A$ hexagon, and let $t_1, \ldots, t_N$ denote the (random) coordinates of N lozenges on the vertical line at distance N from the left edge. The N-dimensional vectors*

$$\left(\frac{t_i - \frac{A}{2}}{\sqrt{\frac{3A}{8}}} \right)_{1 \le i \le N}$$

converge in distribution to the absolutely continuous probability measure on $\{(x_1, \ldots, x_N) \in \mathbb{R}^N : x_1 \le \cdots \le x_N\}$ with density proportional to

$$\prod_{1 \le i < j \le N} (x_i - x_j)^2 \prod_{i=1}^{N} \exp\left(-\frac{x_i^2}{2} \right). \tag{19.5}$$

We can go further and look not only at level N (at distance N from the base) but also at the first N levels simultaneously. In the tilings model, let $t^j = (t^j_1 < \cdots < t^j_j)$ be the positions of horizontal lozenges at the jth level, counted so that we have $0 \le t^j_1 < \cdots < t^j_j \le A + j - 1$. The combinatorics of tilings imply the following interlacing[3] inequalities:

$$t^{j+1}_{i+1} > t^j_i \ge t^{j+1}_i, \text{ for all relevant } i, j.$$

Proposition 19.9 *The $\frac{N(N+1)}{2}$-dimensional vectors*

$$\left(\frac{t^j_i - \frac{A}{2}}{\sqrt{\frac{3A}{8}}} \right)_{1 \le i \le j \le N} \tag{19.6}$$

converge in distribution as $A \to \infty$ to the absolutely continuous probability measure on

$$\{(x^j_i)_{1 \le i \le j \le N} \in \mathbb{R}^{N(N+1)/2} : x^{j+1}_{i+1} \ge x^j_i \ge x^{j+1}_i \text{ for all relevant } i, j\}$$

with density proportional to

$$\prod_{1 \le i < j \le N} (x^N_j - x^N_i) \cdot \prod_{i=1}^{N} \exp\left(-\frac{(x^N_i)^2}{2} \right). \tag{19.7}$$

[3] An interlacing triangular array of integers (or, sometimes, reals) satisfying such inequalities is often called a "Gelfand–Tsetlin pattern." The name originates from the labeling with such patterns of the *Gelfand–Tsetlin basis* in irreducible representations of unitary groups $U(N)$, first constructed in Gel'fand and Tsetlin (1950).

Proof   We set

$$t_i^j = \frac{A}{2} + \sqrt{\frac{3A}{8}} x_i^j.$$

From Proposition 19.8, the law of

$$\left(\frac{t_i^N - \frac{A}{2}}{\sqrt{\frac{3A}{8}}} \right)_{1 \leq i \leq N} \tag{19.8}$$

as $A \to \infty$ has density proportional to

$$\prod_{1 \leq i < j \leq N} (x_j^N - x_i^N)^2 \cdot \prod_{i=1}^N \exp\left(-\frac{(x_i^N)^2}{2} \right) \cdot (1 + o(1)), \quad A \to \infty.$$

By the Gibbs property (following from the uniformity of random tilings), the law of the $\frac{N(N+1)}{2}$–dimensional vectors (19.6) is obtained from the law of $t_1^N < \cdots < t_N^N$, dividing by the number of lozenge tilings of the trapezoid with teeth at the positions t_i^N. Therefore, by Proposition 19.3, we need to multiply the law of the rescaled vector (19.8) by a constant (independent of the t_i^N) multiple of

$$\frac{1}{\prod_{1 \leq i < j \leq N} (t_j^N - t_i^N)} \propto \frac{1}{\prod_{1 \leq i < j \leq N} (x_j^N - x_i^N)}.$$

The result follows.   □

So far, we have only discussed random tilings of hexagons and the scaling limit of their marginals. What is the relation with random matrices, which we promised to explain? In the next lecture we will show that the distribution (19.7) has a random matrix origin: it is called the "GUE-corners process," and it appears in the celebrated Gaussian unitary ensemble.

20

Lecture 20: GUE-Corners Process and Its Discrete Analogues

At the very end of Lecture 19 we obtained an interesting distribution on interlacing arrays of reals $\{x_i^j\}_{1 \le i \le j \le N}$ as a scaling limit for the positions of lozenges in the uniformly random tilings of the hexagon near one of the sides of the hexagon. Our next task is to identify this distribution with the so-called Gaussian unitary ensemble (GUE)-corners process.[1]

20.1 Density of GUE-Corners Process

For a k-dimensional real vector $u = (u_1, \dots, u_k)$ and a $(k+1)$-dimensional real vector $v = (v_1, \dots, v_{k+1})$, we say that u interlaces with v and write $u \prec v$, if

$$v_1 \le u_1 \le v_2 \le u_2 \le \cdots \le u_k \le v_{k+1}.$$

Theorem 20.1 *Let X be an $N \times N$ matrix of independent and identically distributed (i.i.d.) complex Gaussian random variables of the form $N(0,1) + iN(0,1)$, and set $M = \frac{X+X^*}{2}$ to be its Hermitian part. Define $x_1^m \le x_2^m \le \cdots \le x_m^m$ to be the eigenvalues of the principal top-left corner of M, as in Figure 20.1. Then the array $\{x_i^m\}_{1 \le i \le m \le N}$ has the density (with respect to the $\frac{N(N+1)}{2}$-dimensional Lebesgue measure) proportional to*

$$\mathbf{1}_{x^1 < x^2 < \cdots < x^N} \prod_{1 \le i < j \le N} (x_j^N - x_i^N) \prod_{i=1}^N \exp\left(-\frac{(x_i^N)^2}{2}\right). \tag{20.1}$$

---

[1] Another name used in the literature is the "GUE-minors process." Because, by definition, a minor is the determinant of a submatrix (rather than the submatrix itself appearing in the definition of our process), we choose to use the name "GUE-corners process" instead.

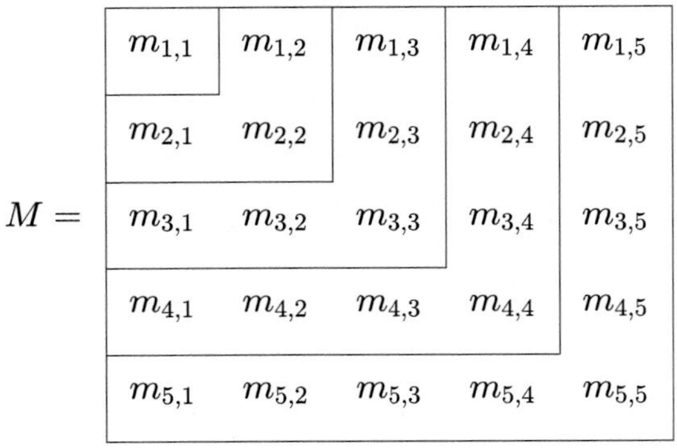

$$M = \begin{array}{|c|c|c|c|c|}\hline m_{1,1} & m_{1,2} & m_{1,3} & m_{1,4} & m_{1,5} \\ m_{2,1} & m_{2,2} & m_{2,3} & m_{2,4} & m_{2,5} \\ m_{3,1} & m_{3,2} & m_{3,3} & m_{3,4} & m_{3,5} \\ m_{4,1} & m_{4,2} & m_{4,3} & m_{4,4} & m_{4,5} \\ m_{5,1} & m_{5,2} & m_{5,3} & m_{5,4} & m_{5,5} \\ \hline \end{array}$$

Figure 20.1 Principal submatrices of M.

Our proof closely follows the exposition of Baryshnikov (2001) and Neretin (2003). One earlier appearance of such arguments is Section 9.3 in Gel'fand and Naimark (1950).

Proof of Theorem 20.1   We proceed by induction in N with the base case $N = 1$ being the computation of the probability density of the Gaussian law.

For the induction step, we need to verify the following conditional density computation:

$$\mathrm{Prob}(x_i^{m+1} = y_i + dy_i, \; 1 \le i \le m + 1 \mid x_i^j, 1 \le i \le j \le m)$$

$$\stackrel{?}{\sim} \mathbf{1}_{x^m \prec y} \frac{\prod_{1 \le i < j \le m+1}(y_j - y_i)}{\prod_{1 \le i < j \le m}(x_j^m - x_i^m)} \exp\left(-\sum_{i=1}^{m+1} \frac{(y_i)^2}{2} + \sum_{i=1}^{m} \frac{(x_i^m)^2}{2}\right), \quad (20.2)$$

where $\sim$ means proportionality up to a coefficient, which depends on m but not on x_i^j or y_i.

Let Y denote the $(m + 1) \times (m + 1)$ corner, and let X denote the $m \times m$ corner. Instead of (20.2), we prove a stronger equality, in which we condition on the full matrix X rather than its eigenvalues. We thus aim to compute the conditional law of the eigenvalues of Y given the corner X:

$$Y = \begin{pmatrix} & & & u_1 \\ & X & & \vdots \\ & & & u_m \\ \bar{u}_1 & \cdots & \bar{u}_m & v \end{pmatrix}.$$

Note that $u_1, \ldots, u_m$ are i.i.d. complex Gaussians, and v is a real Gaussian random variable.

In order to shorten the notations, we use $x_1, \ldots, x_m$ to denote the eigenvalues of X (i.e., $x_i := x_i^m$). Choose a unitary $m \times m$ matrix U, which diagonalizes X (i.e., $UXU^* = \mathrm{diag}(x_1, \ldots, x_m)$). Let $\tilde{U}$ be the $(m+1) \times (m+1)$ unitary matrix with $m \times m$ corner U, the unit diagonal element $U_{m+1,m+1} = 1$, and vanishing remaining elements $U_{m+1,i} = U_{i,m+1} = 0$, $1 \le i \le m$. Set $\tilde{Y} = \tilde{U} Y \tilde{U}^*$, so that

$$\tilde{Y} = \begin{pmatrix} x_1 & & & u_1' \\ & \ddots & & \vdots \\ & & x_m & u_m' \\ \bar{u}_1' & \cdots & \bar{u}_m' & v' \end{pmatrix}.$$

We claim that $u_1', \ldots, u_m'$ are still i.i.d. Gaussians, and v' is still a real Gaussian random variable. Indeed, the law of these random variables is obtained by rotating a Gaussian vector with (independent from it) a unitary matrix, which keeps the distribution unchanged.

Because the conjugation of Y does not change its eigenvalues, it remains to compute the law of the eigenvalues of $\tilde{Y}$. For that, we evaluate the characteristic polynomial of $\tilde{Y}$. Expanding the determinant by the last column, we get

$$\det\left(y \cdot \mathbf{Id} - \tilde{Y}\right) = \prod_{i=1}^m (y - x_i) \cdot \left(y - v' - \sum_{i=1}^m \frac{|u_i'|^2}{y - x_i}\right). \tag{20.3}$$

Let us denote $|u_i'|^2$ through ξ_i. Since u_i' is a two-dimensional Gaussian vector with independent $N(0, 1/2)$ components, ξ_i has a standard exponential distribution of density $\exp(-\xi)$, $\xi > 0$. Because the characteristic polynomial of $\tilde{Y}$ can be alternatively computed as $\prod_{i=1}^{m+1}(y - y_i)$, (20.3) implies

$$\frac{\prod_{i=1}^{m+1}(y - y_i)}{\prod_{i=1}^m (y - x_i)} = y - v' - \sum_{i=1}^m \frac{\xi_i}{y - x_i}. \tag{20.4}$$

Here is a corollary of the last formula:

Lemma 20.2 *We have*

$$\xi_a = -\frac{\prod_{b=1}^{m+1}(x_a - y_b)}{\prod_{1 \le c \le m; c \ne a}(x_a - x_c)}, \quad 1 \le a \le m, \quad and \quad v' = \sum_{b=1}^{m+1} y_b - \sum_{a=1}^m x_a. \tag{20.5}$$

Proof Multiplying (20.4) with $y - x_a$ and then setting $y = x_a$, the formula for ξ_a is obtained. Looking at the coefficient of $\frac{1}{y}$ in the expansion of (20.4) into a series in $\frac{1}{y}$ for large y, we get the formula for v'. $\square$

In addition, we have the following statement, whose proof we leave as an exercise:

Exercise 20.3   In formula (20.4), all the numbers ξ_a, $1 \le a \le m$ are nonnegative if and only if the sequences $\{x_i\}$ and $\{y_i\}$ interlace (i.e., $\vec{x} \prec \vec{y}$).

We continue the proof of Theorem 20.1. Because we know the joint distribution of ξ_a and v', and we know their link to $\{y_i\}$ through (20.4), it only remains to figure out the Jacobian of the change of variables $\{\xi_a$, $1 \le a \le m$; $v'\} \mapsto \{y_i$, $1 \le i \le m+1\}$. Straightforward computation based on Lemma 20.2 shows that

$$\frac{\partial \xi_a}{\partial y_b} = \xi_a \cdot \frac{1}{x_a - y_b}, \qquad \frac{\partial v'}{\partial y_b} = 1.$$

Hence, the desired Jacobian is

$$\xi_1 \cdots \xi_m \cdot \det \begin{pmatrix} & & 1 \\ & \frac{1}{x_a - y_b} & \vdots \\ & & 1 \end{pmatrix}. \tag{20.6}$$

We can simplify the last formula using the following determinant computation:

Lemma 20.4 (Cauchy determinant formula)   *For any* $u_1, \ldots, u_N$, $v_1, \ldots, v_N$, *we have*

$$\det \left(\frac{1}{u_a - v_b} \right)_{a,b=1}^{N} = \frac{\prod_{a<a'}(u_a - u_{a'}) \prod_{b<b'}(v_{b'} - v_b)}{\prod_{a,b=1}^{N}(u_a - v_b)}. \tag{20.7}$$

Proof   Multiply (20.7) by $\prod_{a,b=1}^{N}(u_a - v_b)$ and notice that the left-hand side becomes a polynomial in u_i and v_j of degree $N(N-1)$. This polynomial is skew symmetric both in variables $(u_1, \ldots, u_N)$ and in variables $(v_1, \ldots, v_N)$. Therefore, it has to be divisible by $\prod_{a<a'}(u_a - u_{a'}) \prod_{b<b'}(v_{b'} - v_b)$. Comparing the degrees, we conclude that (20.7) has to hold up to multiplication by a constant. Comparing the coefficients of the leading monomials, we see that this constant is one.   □

Sending one of the variables in (20.7) to infinity, we compute the determinant in (20.6). Hence, the desired Jacobian is

$$\xi_1 \cdots \xi_m \cdot \frac{\displaystyle\prod_{1 \le a < a' \le m}(x_a - x_{a'}) \prod_{1 \le b < b' \le m+1}(y_b - y_{b'})}{\displaystyle\prod_{a=1}^{m}\prod_{b=1}^{m+1}(x_a - y_b)} = \frac{\displaystyle\prod_{1 \le b < b' \le m+1}(y_b - y_{b'})}{\displaystyle\prod_{1 \le a < a' \le m}(x_a - x_{a'})}.$$

Using the explicit formula for the joint density of $\{\xi_a, 1 \le a \le m; v'\}$, we conclude that the conditional density of $y_1, \ldots, y_{m+1}$ given $x_1, \ldots, x_m$ is proportional to

$$\frac{\displaystyle\prod_{1 \le b < b' \le m+1} (y_b - y_{b'})}{\displaystyle\prod_{1 \le a < a' \le m} (x_a - x_{a'})} \exp\left(-\sum_{a=1}^{m} \xi_a - \frac{(v')^2}{2} \right).$$

Inserting the indicator function from Exercise 20.3 into the last formula, we get (20.2), and we notice that the definition of the matrix $\tilde{Y}$ implies

$$\sum_{a=1}^{m} \xi_a + \frac{(v')^2}{2} = \frac{1}{2}\left(\text{Trace}(\tilde{Y}^2) - \sum_{a=1}^{m} (x_a)^2 \right) = \frac{1}{2}\left(\sum_{b=1}^{m+1} (y_b)^2 - \sum_{a=1}^{m} (x_a)^2 \right). \quad \Box$$

Exercise 20.5 Repeat the argument for *real* (rather than complex) matrix X with i.i.d. $N(0, 1)$ matrix elements.[2] How does the density (20.1) change?

20.2 GUE-Corners Process as a Universal Limit

In Lecture 19 we obtained the GUE-corners process as a scaling limit for the random lozenge tilings of hexagons. In fact, it appears much wider.

Conjecture 20.6 (Okounkov and Reshetikhin, 2006; Johansson and Nordenstam, 2006) *The GUE-corners process is a universal scaling limit for uniformly random lozenge tilings at the points where the boundary of the frozen boundary of a tiling is tangent to a straight segment of the boundary of the tiled domain.*

For trapezoid domains Conjecture 20.6 was settled in Gorin and Panova (2015). For almost general domains it was recently proven in Aggarwal and Gorin (2021).

Similarly to the extended sine process (or ergodic Gibbs translation-invariant measure), which we saw in the bulk of the tilings, and the extended Airy process (or Airy line ensemble) that appears at the edges of liquid regions, the GUE-corners process can be obtained as a solution to a certain classification problem. That was an argument of Okounkov and Reshetikhin (2006) for the validity of Conjecture 20.6. The classification problem in which the GUE-corners process appears can be linked to the representation theory and to the ergodic theory, and we now present it.

Let $U(N)$ denote the group of all (complex) $N \times N$ unitary matrices. There is a natural embedding of $U(N)$ into $U(N + 1)$ identifying an operator in

---

[2] In this case $M = \frac{X + X^*}{2}$ is known as the "Gaussian orthogonal ensemble."

the N-dimensional space with a one in the $N + 1$-dimensional space fixing the last basis vector. Through these embeddings, we can define the infinite-dimensional unitary group as the union of all finite-dimensional groups: $U(\infty) = \bigcup_{N=1}^{\infty} U(N)$. Each element of $U(\infty)$ is an infinite matrix whose upper-left $k \times k$ corner is a unitary matrix of size k, and the rest of the matrix is filled with 1's on the diagonal and 0's everywhere else.

Consider the space $\mathcal{H}$ of all infinite complex Hermitian matrices. Their matrix elements $h_{ij} \in \mathbb{C}$, $i, j = 1, 2, 3, \ldots$ satisfy $\overline{h_{ij}} = h_{ji}$. The group $U(\infty)$ acts on $\mathcal{H}$ by conjugations, and we can study $U(\infty)$-invariant probability measures on $\mathcal{H}$. We call such a measure "ergodic" if it is an extreme point of the convex set of all $U(\infty)$-invariant probability measures.

One can construct three basic examples of ergodic measures or, equivalently, infinite random Hermitian matrices:

1. A deterministic **diagonal matrix:** $\gamma_1 \cdot \text{Id}$, $\gamma_1 \in \mathbb{R}$.
2. **A GUE:** Take $\gamma_2 \in \mathbb{R}_{>0}$ and an infinite matrix X of i.i.d. $N(0, 1) + \mathbf{i}N(0, 1)$ matrix elements. Consider a random matrix $\gamma_2 \frac{X+X^*}{2}$.
3. **Wishart (or Laguerre) ensemble:** Take $\alpha \in \mathbb{R}$ and an infinite vector V of i.i.d. $N(0, 1) + \mathbf{i}N(0, 1)$ components. Consider a rank 1 random matrix $\alpha V V^*$.

A theorem of Pickrell (1991) and Olshanski and Vershik (1996) claims that the sums of independent matrices from these three ensembles exhaust the list of ergodic $U(\infty)$-invariant measures.

Theorem 20.7 (Pickrell, 1991; Olshanski and Vershik, 1996)   *Ergodic $U(\infty)$-invariant measures on $\mathcal{H}$ are parameterized by $\gamma_1 \in \mathbb{R}$, $\gamma_2 \in \mathbb{R}_{\geq 0}$ and a collection of real numbers $\{\alpha_i\}$, such that $\sum_i \alpha_i^2 < \infty$. The corresponding random matrix is given by*

$$\gamma_1 \cdot \text{Id} + \gamma_2 \frac{X + X^*}{2} + \sum_i \alpha_i \left(\frac{V_i V_i^*}{2} - \text{Id} \right), \tag{20.8}$$

where all terms are taken to be independent.

Note that, in principle, the subtraction of $\sum_i \alpha_i \text{Id}$ could have been absorbed into the γ_1 term. However, in this case the sum over i would have failed to converge (on the diagonal) for sequences α_i, such that $\sum_i \alpha_i^2 < \infty$, but $\sum_i |\alpha_i| = \infty$.

One sees that the GUE-corners process appears in a particular instance of Theorem 20.7 with $\gamma_1 = 0$ and $\alpha_i = 0$. Let us explain how to link this appearance to the limits of random tilings.

Lemma 20.8 *Let $M \in \mathcal{H}$ be a random $U(\infty)$-invariant infinite Hermitian matrix, and let $m_1^N \leq m_2^N \leq \cdots \leq m_N^N$ denote the eigenvalues of its principle $N \times N$ top-left corner. Then the eigenvalues m_i^j interlace, which means*

$$m_i^{N+1} \leq m_i^N \leq m_{i+1}^{N+1}, \quad 1 \leq i \leq N.$$

Moreover, they satisfy the Gibbs *property, which says that conditionally on the values of $m_1^N, m_2^N, \ldots, m_N^N$, the $\frac{N(N-1)}{2}$ eigenvalues m_i^j, $1 \leq i \leq j < N$ have a uniform distribution on the polytope determined by the interlacing conditions.*

Sketch of the Proof The interlacement is a general linear algebra fact about the eigenvalues of the corners of a Hermitian matrix. The conditional uniformity can be proven by a version of the argument that we used in determining the density of the GUE-corners process in Theorem 20.1. □

In fact, one can show that the classification of ergodic $U(\infty)$-invariant random Hermitian matrices of Theorem 20.7 is equivalent to the classification of all ergodic Gibbs measures on infinite arrays m_j^i of interlacing real numbers.

Lemma 20.8 explains why we expect the scaling limit of the uniform measure on tilings near a straight boundary of the domain to be one of the measures in Theorem 20.1: horizontal lozenges in tilings near the boundary satisfy the discrete versions of the interlacement and Gibbs properties. Hence, we expect continuous analogues to hold for the limiting object. But why do we observe only the GUE-corners process but no deterministic or Wishart components?

This can be explained by the very different law of large numbers for these ensembles. Take a random infinite Hermitian matrix M, and consider the empirical measure of the eigenvalues of its corner:

$$\delta^N = \frac{1}{N} \sum_{i=1}^{N} \delta_{m_i^N}.$$

The $N \to \infty$ behavior of the empirical measure is very different for the three basic examples:

1. For the diagonal matrix $\gamma_1 \cdot \mathrm{Id}$, δ^N is a unit mass at γ_1 for all N.
2. For the GUE ensemble with $\gamma_2 = 1$, the eigenvalues as $N \to \infty$ fill the segment $[-2\sqrt{N}, 2\sqrt{N}]$; after normalization by $\sqrt{N}$, the measure δ^N converges to the *Wigner semicircle law*.
3. For the rank 1 Wishart matrix $\frac{1}{2}VV^*$, the empirical measures δ^N has mass $\frac{N-1}{N}$ at 0 with an outlier of mass $\frac{1}{N}$ at point $\approx N$ as $N \to \infty$.

Exercise 20.9 Prove the last statement about rank 1 Wishart matrices.

On the other hand, in the tilings picture, we deal with a point where the frozen boundary is *tangent* to the straight boundary of the domain. Hence, the frozen boundary locally looks like a parabola, which is consistent only with a $[-2\sqrt{N}, 2\sqrt{N}]$ segment filled with eigenvalues in the GUE case.

Conceptual Conclusion. The GUE-corners process is expected to appear in the scaling limit near the frozen-boundary tangency points whenever we deal with interlacing triangular arrays of particles satisfying (asymptotically) conditional uniformity.

This prediction matches rigorous results going beyond random tilings. For instance, Gorin and Panova (2015), Gorin (2014), and Dimitrov (2020) demonstrated the appearance of the GUE-corners process as the scaling limit of the six-vertex model.

20.3 A Link to Asymptotic Representation Theory and Analysis

In Theorem 20.7 and Lemma 20.8 we saw how the GUE-corners process gets linked to the classification problem of ergodic theory or asymptotic representation theory. But in fact, the same connection already exists on the discrete level of random lozenge tilings.

Let us present a discrete-space analogue of Theorem 20.7. For that, we let $\mathcal{T}$ denote the set of all lozenge tilings of the (right) half-plane, such that far up, there are only lozenges of type $\lozenge$ and far down there are only lozenges of type $\lozenge$, see Figure 20.2. This constraint implies that there is precisely one horizontal lozenge on the first vertical line, precisely two on the second one, precisely three on the third one, and so forth. Their coordinates interlace, as in Section 10.3 or in Proposition 19.9. When one forgets about tilings, considering only these coordinates, then the resulting object is usually called an (infinite) "Gelfand–Tsetlin pattern" or a "path in the Gelfand–Tsetlin graph."

We aim to study the probability measures on $\mathcal{T}$, which are Gibbs in the same sense as in Definition 13.4 of Lecture 13. Because of our choice of boundary conditions, the interlacing can be equivalently restated in terms of horizontal lozenges $\diamond$: given the positions of N horizontal lozenges on the Nth vertical, the conditional distribution of $N(N-1)/2$ horizontal lozenges on the first $N-1$ verticals is uniform on the set defined by the interlacing conditions.

Theorem 20.10 (Aissen et al., 1952; Edrei, 1953; Voiculescu, 1976; Vershik and Kerov, 1982; Borodin and Olshanski, 2012; Petrov, 2014b; Gorin and

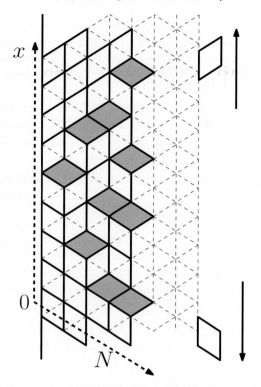

Figure 20.2   The first four verticals of a tiling of the right half-plane, which has prescribed lozenges far up and far down. Horizontal lozenges interlace and are shown in dark gray. The coordinate system is shown by dashed lines.

Panova, 2015)   *The ergodic Gibbs measures on $\mathcal{T}$ are parameterized by two reals $\gamma^+ \geq 0$, $\gamma^- \geq 0$ and four sequences of nonnegative reals:*

$$\alpha_1^+ \geq \alpha_2^+ \geq \cdots \geq 0, \quad \alpha_1^- \geq \alpha_2^- \geq \cdots \geq 0,$$

$$\beta_1^+ \geq \beta_2^+ \geq \cdots \geq 0, \quad \beta_1^- \geq \beta_2^- \geq \cdots \geq 0,$$

such that $\sum_{i=1}^{\infty} (\alpha_i^+ + \beta_i^+ + \alpha_i^- + \beta_i^-) < \infty$ and $\beta_1^+ + \beta_1^- \leq 1$.

One exciting feature of Theorem 20.10 (which is also a reason for so many authors producing different proofs of it) is that it is equivalent to two other important theorems, one from the representation theory and another one from the classical analysis. Let us present them.

Recall that infinite-dimensional unitary group $U(\infty)$ is defined as the inductive limit or union of finite-dimensional unitary groups $U(1) \subset U(2) \subset$

$U(3) \subset \cdots \subset U(\infty)$. A (normalized) *character* $\chi(u)$ of $U(\infty)$ is a continuous complex function on the group, which is as follows:

1. *Central*, that is, $\chi(ab) = \chi(ba)$ for all $a, b \in U(\infty)$;
2. *Positive-definite*, that is, for each $k = 1, 2, \ldots$, complex numbers $z_1, \ldots, z_k$, and elements of the group $u_1, \ldots, u_k$, we have

$$\sum_{i,j=1}^{k} z_i \bar{z}_j \chi(u_i u_j^{-1}) \geq 0; \quad \text{and}$$

3. *Normalized*, that is, $\chi(e) = 1$, where e is the unit element (diagonal matrix of 1's) in $U(\infty)$.

The characters of $U(\infty)$ form a convex set; that is, the convex linear combination of characters is again a character, and we are interested in *extreme* ones, which cannot be decomposed into a linear combination of others. As it turns out, the complete list of extreme characters is given by Theorem 20.10. The value of an extreme character corresponding to the parameters of Theorem 20.10 on the matrix $U \in U(\infty)$ is given by

$$\chi(U) = \prod_{u \in \text{Spectrum}(U)} e^{\gamma^+(u-1)+\gamma^-(u^{-1}-1)} \prod_{i=1}^{\infty} \frac{1+\beta_i^+(u-1)}{1-\alpha_i^+(u-1)} \cdot \frac{1+\beta_i^-(u^{-1}-1)}{1-\alpha_i^-(u^{-1}-1)}.$$
(20.9)

Note that all but finitely many eigenvalues of a matrix $U \in U(\infty)$ are equal to one; therefore, the product over u in (20.9) is actually finite. The double multiplicativity (in u and in i) of the extreme characters is a remarkable feature of $U(\infty)$, which has analogues for other *infinite-dimensional* groups. However, nothing like this is true for the finite-dimensional unitary groups $U(N)$, whose extreme characters are much more complicated (normalized) Schur polynomials $s_\lambda(u_1, \ldots, u_N)$ (see Definition 22.3 in Lecture 22).

The correspondence between Theorem 20.10 and the characters of $U(\infty)$ goes through the expansion of the restrictions of the characters. Take *any* character χ of $U(\infty)$, and consider its restriction to $U(N)$. This restriction is a symmetric function in N eigenvalues $u_1, \ldots, u_N$ of a $N \times N$ unitary matrix and can be represented as a line combination of Schur polynomials as follows:

$$\chi|_{U(N)}(u_1, \ldots, u_N) = \sum_{\lambda=(\lambda_1 \geq \lambda_2 \geq \cdots \geq \lambda_N) \in \mathbb{Z}^N} P^N(\lambda) \frac{s_\lambda(u_1, \ldots, u_N)}{s_\lambda(1, \ldots, 1)}. \quad (20.10)$$

We can prove that the coefficients $P^N(\lambda)$ are nonnegative and sum up to one: the first condition follows from positive-definiteness, and the second one is implied by the normalization $\chi(e) = 1$. By setting $\mu_i = \lambda_i + N - i$, these coefficients thus define a probability distribution on N tuples of horizontal lozenges as at the Nth vertical line of the lozenge tiling of the right half-plane; see Figure 20.2. With a bit more work, we can prove that the correspondence $\chi \leftrightarrow \{P^1, P^2, P^3, \ldots\}$ of (20.10) is a bijection between the characters of $U(\infty)$ and the Gibbs measures on $\mathcal{T}$.

For another equivalent form of Theorem 20.10, consider a doubly infinite sequence of reals $\ldots d_{-2}, d_{-1}, d_0, d_1, d_2, \ldots$ and introduce a doubly infinite Toeplitz matrix D by $D[i, j] = d_{j-i}$, $i, j = \ldots, -1, 0, 1, \ldots$. The matrix D is called "totally positive" if all (finite) *minors* of D are nonnegative. For a normalization, we further assume that the sum of elements in each row of D is one (i.e., $\sum_{i=-\infty}^{\infty} d_i = 1$).[3] Then the classification of totally positive doubly infinite Toeplitz matrices is equivalent to Theorem 20.10, with bijection given by the generating series (convergent for u in the unit circle on complex plane):

$$\sum_{i=-\infty}^{\infty} u^i d_i = e^{\gamma^+(u-1)+\gamma^-(u^{-1}-1)} \prod_{i=1}^{\infty} \frac{1 + \beta_i^+(u-1)}{1 - \alpha_i^+(u-1)} \cdot \frac{1 + \beta_i^-(u^{-1}-1)}{1 - \alpha_i^-(u^{-1}-1)}.$$

The classification of totally positive Toeplitz matrices was originally motivated by a data-smoothing problem. Suppose that we have a real data sequence x_i, $i \in \mathbb{Z}$ and a pattern d_i, $i \in \mathbb{Z}$; then we can construct a smoothed-data sequence y_i through

$$y_i = \sum_j x_{i-j} d_j.$$

We can show that the number of sign changes ("oscillations") in y_i is always smaller than that in x_i if and only if the corresponding Toeplitz matrix $D[i, j] = d_{j-i}$ is totally positive.

The reason why Theorem 20.10 is linked to totally positive Toeplitz matrices lies in the formulas for the coefficients in the decomposition (20.10). On the one hand, they must be nonnegative because they define probability measures, but on the other hand, they are essentially the minors of a certain Toeplitz matrix, as the following exercise shows:

Exercise 20.11   Consider the following decomposition of a symmetric two-sided power series in variables $u_1, \ldots, u_N$:

$$\prod_{i=1}^{N} \left(\sum_{k \in \mathbb{Z}} c_k (u_i)^k \right) = \sum_{\lambda = (\lambda_1 \geq \lambda_2 \geq \cdots \geq \lambda_N) \in \mathbb{Z}^N} c_\lambda \cdot \frac{\det \left[u_i^{\lambda_j + N - j} \right]_{i,j=1}^{N}}{\prod_{1 \leq i < j \leq N} (u_i - u_j)}. \quad (20.11)$$

---

[3] A relatively simple argument shows that if D is a totally positive Toeplitz matrix, then either $d_i = CR^i$ for all i, or replacing d_i by $CR^i d_i$ with appropriate C and R, we can guarantee the condition $\sum_{i=-\infty}^{\infty} d_i = 1$. Therefore, the normalization condition does not reduce the generality.

Multiplying both sides by the denominator, show that the coefficients c_λ are computed as

$$c_\lambda = \det \left[c_{\lambda_i - i + j} \right]_{i,j=1}^N.$$

Using the *Weyl character formula* (cf. Weyl, 1939), one identifies the Schur functions $s_\lambda(u_1, \ldots, u_N)$ with the ratios on the right-hand side of (20.11). Hence, the coefficients c_λ of Exercise 20.11 differ from $P^N(\lambda)$ in (20.10) by the factor $s_\lambda(1, \ldots, 1)$. This factor is a positive integer (it counts the dimension of a representation of $U(N)$ as well as the lozenge tilings of a trapezoid domain; see Proposition 19.3 and Section 22.2) – hence, the nonnegativity of c_λ and $P^N(\lambda)$ is the same condition.

The character theory for $U(\infty)$ also admits a q-deformation, which turns out to be closely related to the study of both q^{Volume}-weighted random tilings and *quantum groups*; see Gorin (2012), Gorin and Olshanski (2016), and Sato (2019).

We end this lecture with the remark that both Theorem 20.7 and Theorem 20.10, as well as their q-deformations, fit into a general framework for studying *Gibbs measures on branching graphs*. These graphs can be very different, and we refer the reader to Kerov (2003), Borodin and Olshanski (2013), and Section 7 of Borodin and Olshanski (2016) for some details and examples.

21

Lecture 21: Discrete Log-Gases

21.1 Log-Gases and Loop Equations

Recall that in Lecture 19 we computed in (19.1) the law of horizontal lozenges on the Nth vertical of a randomly tiled equilateral hexagon with side length A:

$$\mathbb{P}(x_1, \ldots, x_N) \propto \prod_{1 \leq i < j \leq N} (x_i - x_j)^2 \prod_{i=1}^{N} w(x_i), \qquad (21.1)$$

where $w(x) = (x + 1)_{A-N}(A + N - x)_{A-N}$. The distribution of form (21.1) with arbitrary w is usually called "log-gas" in statistical mechanics. The name originates from the traditional rewriting of (21.1) as

$$\exp\left(\beta \sum_{1 \leq i < j \leq N} H(x_i, x_j) + \sum_{i=1}^{N} \ln w(x_i) \right), \qquad H(x, y) = \ln |x-y|, \qquad \beta = 2,$$
$$(21.2)$$

where $H(x, y)$ represents the functional form of the interaction between particles (which is logarithmic in our case), and β is the strength of the interaction.

The distributions of the form (21.2) with $x_i \in \mathbb{R}$ or $x_i \in \mathbb{C}$ are widespread in random-matrix theory, where they appear as distributions of eigenvalues for various ensembles of matrices; see Forrester (2010). In this setting, the parameter β is the dimension of the (skew) field to which the matrix elements belong. Here $\beta = 1, 2, 4$ corresponds to reals, complex numbers, and quaternions, respectively. General real values of parameter β also have relevance in theoretical physics and statistical mechanics, but we are not going to discuss this here.

One powerful tool for the study of log-gases as the number of particles N goes to infinity is based on certain exact relations for the expectation of observables[1]

---

[1] By an observable, we mean a function $f(X)$ of the random system state X, such that the expectation $\mathbb{E} f(X)$ is accessible either through explicit formulas or equations fixing it.

in the system. The usefulness of these relations has been understood by the mathematical community since Johansson (1998); before that, these relations were known to theoretical physicists under the name of "loop equations" or "Dyson–Schwinger equations"; we refer the reader to Guionnet (2019) for a recent review.

These equations can be treated as far-reaching generalizations of the following characterization of the Gaussian law. Let ξ be a standard normal $N(0, 1)$ random variable, and let $f(x)$ be a polynomial. Then straightforward integration by parts shows that

$$\mathbb{E}[\xi f(\xi)] = \mathbb{E}[f'(\xi)], \qquad (21.3)$$

and the relation (21.3) can be used to compute all the moments $\mathbb{E}\,\xi^k$ and, hence, to reconstruct the distribution of ξ.

In a similar spirit, the loop equations for (continuous-space) log-gases can be obtained by a smart integration by parts and further used to extract probabilistic information.

In the context of this book, we are interested in the discrete setting when the particles x_i in (21.1) are constrained to the lattice $\mathbb{Z}$. It took a while to generalize the loop equations to such a setting because the naive replacement of integration by parts with summation by parts did not seem to work. The correct form of the equations was recently introduced in Borodin et al. (2017); it was guided by the notion of qq-characters of Nekrasov (2016).

Theorem 21.1 (Nekrasov Equation, Discrete Loop Equation)   *Consider a point process* $(x_1 < x_2 < \cdots < x_N) \subset \{0, \ldots, M\}$ *with a law of the form* (21.1). *Suppose that there exist functions* $\varphi^+(z), \varphi^-(z)$ *such that the following hold:*

1. *Functions* $\varphi^+(z)$ *and* $\varphi^-(z)$ *are holomorphic in a complex neighborhood of* $[0, M + 1]$.
2. *For any* $x \in \{1, \ldots, M\}$, *we have*

$$\frac{w(x)}{w(x - 1)} = \frac{\varphi^+(x)}{\varphi^-(x)}.$$

3. $\varphi^+(M + 1) = \varphi^-(0) = 0$.

Then the function $R_N(z)$ *defined by*

$$R_N(z) = \varphi^-(z)\,\mathbb{E} \prod_{i=1}^{N} \left(1 - \frac{1}{z - x_i}\right) + \varphi^+(z)\,\mathbb{E} \prod_{i=1}^{N} \left(1 + \frac{1}{z - x_i - 1}\right) \qquad (21.4)$$

is holomorphic in the neighborhood of $[0, M + 1]$.

Remark 21.2   The theorem admits a generalization to distributions involving an additional parameter β similarly to (21.2). However, rather than the straightforward $\prod_{i<j} |x_i - x_j|^\beta$, one has to use a more delicate deformation involving gamma functions; see Borodin et al. (2017) for the details.

Proof of Theorem 21.1   Note that $R_N(z)$ is a meromorphic function in the neighborhood of $[0, M]$ with possible poles only at $\{0, 1, \ldots, M + 1\}$. More precisely, the first term of (21.4) might have poles only at $z = x_i \in \{0, \ldots, M\}$ and the second term only at $z = x_i + 1 \in \{1, \ldots, M + 1\}$.

Thus, in order to show that $R_N(z)$ is holomorphic, it is enough to prove that $R_N(z)$ has no poles at $\{0, \ldots, M + 1\}$. Note that in the point process $(x_1, \ldots, x_N)$, all particles are almost surely at distinct positions; hence, all poles of R_N must be simple.

For the point $z = 0$, we have $\varphi^-(0) = 0$; hence, in the first term of (21.4), a possible simple pole at $z = 0$ cancels out with the zero of φ^-, whereas the second term has no pole at 0. Therefore, $R_N(z)$ has no pole at 0. Similarly, for $z = M + 1$, a possible pole in the second term cancels out, whereas the first term has no poles. Thus, $R_N(z)$ has no pole at $M + 1$.

Let $m \in \{1, \ldots, M\}$. Because all poles of $R_N(z)$ are simple, it is enough to prove that the residue at m vanishes. Residue at m results from configurations with $x_i = m$ for some i (first term) or configurations with $x_i = m + 1$ for some i. Expanding $\mathbb{E}$ in (21.4), we get

$$\mathrm{res}_m R_N = \sum_i \mathrm{res}_m^{(i)} R_N,$$

where

$$\mathrm{res}_m^{(i)} R_N = - \sum_{\vec{x}:x_i=m} \varphi^-(m) \mathbb{P}(\vec{x}) \prod_{j \neq i} \left(1 - \frac{1}{m - x_j}\right)$$

$$+ \sum_{\vec{x}:x_i=m-1} \varphi^+(m) \mathbb{P}(\vec{x}) \prod_{j \neq i} \left(1 + \frac{1}{m - x_j - 1}\right).$$

Note that we can pair configurations $\vec{x}$ in the first term and $\vec{x}^{-(i)} = (x_1, \ldots, x_i - 1, \ldots, x_N)$ in the second term. The poles cancel out, leading to $\mathrm{res}_m^{(i)} R_N = 0$, once we prove

$$\varphi^-(m)\mathbb{P}(x_1,\ldots,x_{i-1},m,x_{i+1},\ldots,x_N)\prod_{j\neq i}\left(1-\frac{1}{m-x_j}\right)$$

$$=\varphi^+(m)\mathbb{P}(x_1,\ldots,x_{i-1},m-1,x_{i+1},\ldots,x_N)\prod_{j\neq i}\left(1+\frac{1}{m-x_j-1}\right).$$

$$(21.5)$$

The validity of (21.5) can be checked directly using (21.1):

$$\text{LHS} = \varphi^-(m)w(m)\prod_{\substack{j,k\neq i\\j<k}}(x_j-x_k)^2\prod_{j\neq i}\left[(m-x_j)^2w(x_j)\frac{m-x_j-1}{m-x_j}\right]$$

$$=\varphi^+(m)w(m-1)\prod_{\substack{j,k\neq i\\j<k}}(x_j-x_k)^2\prod_{j\neq i}\left[(m-x_j-1)^2w(x_j)\frac{m-x_j}{m-x_j-1}\right]$$

$$= \text{RHS.} \qquad\qquad\qquad\qquad\qquad\qquad\qquad\qquad\qquad\qquad\qquad \square$$

Remark 21.3   For this cancellation argument, we also should check the boundary cases when one of the configurations $\vec{x}, \vec{x}^{-(i)}$ is not well defined. But the previous computation shows that in such cases, the corresponding term in $\text{res}_m^{(i)}R_N$ will vanish either due to $\prod(x_i-x_j)^2$ in the law or due to the boundary condition $\varphi^+(M+1)=\varphi^-(0)=0$.

Theorem 21.1 can be used to describe various probabilistic properties of log-gases. In particular, it can be applied to the asymptotic study of the uniformly random lozenge tilings of the hexagon. We are going to show two applications. The first one produces another way (in addition to the approaches of Lectures 10 and 16) to compute the limit shape for the tilings and see the frozen-boundary – the inscribed ellipse. The second one (we only sketch the arguments there) leads to the Gaussian fluctuations, which we already expect to be described by the Gaussian free field after Lectures 11 and 12.

21.2 Law of Large Numbers through Loop Equations

We start by specializing Theorem 21.1 to the distribution of the lozenges on the section of a hexagon, as in (19.1), (21.1).

Note that for $w(x)=(x+1)_{A-N}(A+N-x)_{A-N}$, we have

$$\frac{w(x)}{w(x-1)}=\frac{x+A-N}{x}\frac{A+N-x}{2A-x}.$$

Hence, we can take

$$\varphi^+(z) = (z + A - N)(A + N - z), \qquad \varphi^-(z) = z(2A - z),$$

and by Theorem 21.1, the function

$$
\frac{R_N(Nz)}{N^2} = z\left(\frac{2A}{N} - z\right)\mathbb{E}\prod_{i=1}^{N}\left(1 - \frac{1}{N(z - \frac{x_i}{N})}\right)
$$

$$
+ \left(z + \frac{A - N}{N}\right)\left(\frac{A + N}{N} - z\right)\mathbb{E}\prod_{i=1}^{N}\left(1 + \frac{1}{N(z - \frac{x_i}{N} - \frac{1}{N})}\right)
$$

is an entire function. For the limit shape of tilings, we want to consider $N \to \infty$ with $A = \tilde{a}N$ for some fixed $\tilde{a}$. Note that for z away from $[0, \frac{A+N}{N}]$, we have

$$
\prod_{i=1}^{N}\left(1 - \frac{1}{N(z - \frac{x_i}{N})}\right) = \exp\left(\sum_{i=1}^{N}\ln\left(1 - \frac{1}{N(z - \frac{x_i}{N})}\right)\right)
$$

$$
= \exp\left(-\frac{1}{N}\sum_{i}\frac{1}{z - \frac{x_i}{N}} + O(1/N)\right) \to \exp(-G(z)),
$$

where $G(z)$ is the Cauchy–Stieltjes transform:

$$
G(z) = \int_0^{\tilde{a}+1}\frac{1}{z - y}\mu(y)\,dy,
$$

and μ is the limiting density at $(1, y)$ of horizontal lozenges $\diamond$ as $N \to \infty$, similarly to Section 10.3. Recall that the existence of the deterministic μ (or, more generally, the limit shape) was proved earlier using either a concentration of martingales or a variational problem (see Lectures 5–7).

Repeating the same computation for the second term, for z away from $[0, \tilde{a} + 1]$, we have

$$
\frac{R_N(Nz)}{N^2} \xrightarrow[N\to\infty]{} R_\infty(z) = z(2\tilde{a}-z)\exp(-G(z))+(z+\tilde{a}-1)(\tilde{a}+1-z)\exp(G(z)).
$$
$$(21.6)$$

Proposition 21.4 $R_\infty(z)$ *is an entire function.*

Proof Note that outside of $[0; \tilde{a} + 1]$, the holomorphicity is immediate. Moreover, outside of $[0; \tilde{a} + 1]$, entire functions $\frac{R_N(Nz)}{N^2}$ converge to $R_\infty(z)$.

Take a closed contour C around $[0; \tilde{a} + 1]$. For any z inside C, the Cauchy integral formula reads as follows:

$$
\frac{R_N(Nz)}{N^2} = \frac{1}{2\pi i}\oint_C\frac{R_N(Nv)}{N^2(v - z)}\,dv.
$$

Taking $N \to \infty$, we get a holomorphic inside C function:

$$\frac{1}{2\pi \mathbf{i}} \oint_C \frac{R_\infty(v)}{(v-z)} dv,$$

which coincides with $R_\infty(z)$ outside of $[0; \tilde{a}+1]$. Hence $R_\infty(z)$ is entire. □

Corollary 21.5 $R_\infty(z)$ *is a degree two polynomial.*

Proof R_∞ is an entire function growing as z^2 when $z \to \infty$. Hence, R_∞ is a degree two polynomial by Liouville's theorem (by Cauchy's integral formula, the second derivative is bounded and hence constant). □

Let us explicitly compute this degree two polynomial. Note that

$$G(z) = \frac{1}{z} + \frac{c}{z^2} + \cdots$$

when $z \to \infty$. Hence, using Taylor expansion for exp, for $z \to \infty$, we have

$$R_\infty(z) = z(2\tilde{a}-z)\left(1 - G(z) + \frac{G(z)^2}{2} + o\left(z^{-2}\right)\right)$$

$$+ (z+\tilde{a}-1)(\tilde{a}+1-z)\left(1 + G(z) + \frac{G(z)^2}{2} + o\left(z^{-2}\right)\right)$$

$$= \left(-z^2 + (2\tilde{a}+1)z + \left(c - 2\tilde{a} - \tfrac{1}{2}\right)\right) + \left(-z^2 + (2-1)z + \left(\tilde{a}^2 - 1 - c + 2 - \tfrac{1}{2}\right)\right)$$

$$+ o(1) = -2z^2 + 2(\tilde{a}+1)z + \tilde{a}^2 - 2\tilde{a} + o(1).$$

But $R_\infty(z)$ is a polynomial; hence, the $o(1)$ term is equal to zero, and

$$R_\infty(z) = -2z^2 + 2(\tilde{a}+1)z + \tilde{a}^2 - 2\tilde{a}.$$

Comparing the definition (21.6) of R_∞ with the previous expression, we get a quadratic equation for $\exp(G(z))$, namely:

$$z(2\tilde{a}-z)\exp(-G(z)) + (z+\tilde{a}-1)(\tilde{a}+1-z)\exp(G(z))$$
$$= -2z^2 + 2(\tilde{a}+1)z + \tilde{a}^2 - 2\tilde{a}. \quad (21.7)$$

Solving it, we can get an explicit formula for $G(z)$, which is a Stieltjes transform of μ; hence, we can recover μ. Integrating it, we get a formula for the limit shape (asymptotic height function of tilings). Therefore, Theorem 21.1 gives one more approach to computing the law of large numbers (LLN), with others being the Kenyon–Okounkov theory and steepest-descent analysis of the correlation kernel.

Exercise 21.6 For each fixed $\tilde{a}$, the function $G(z)$ found from (21.7) has two real branching points (of type $\sqrt{z}$; they correspond to the points where the density of μ reaches 0 or 1). If we start varying $\hat{a}$, then these branching points form a curve in the $(\frac{1}{\hat{a}}, z)$ plane. Find out how this curve corresponds to the circle inscribed into the unit hexagon.

Hint: Dealing with lozenges along the vertical line at a distance of 1 from the left boundary of a $\tilde{a} \times \tilde{a} \times \tilde{a}$ hexagon is essentially the same as dealing with lozenges along the vertical line at a distance of $\frac{1}{\hat{a}}$ from the left boundary of $1 \times 1 \times 1$ hexagon.

Exercise 21.7 Consider the distribution on $0 \le x_1 < x_2 < \cdots < x_N \le M$ with weight proportional to[2]

$$\prod_{1 \le i < j \le N} (x_i - x_j)^2 \prod_{i=1}^N \binom{M}{x_i}. \tag{21.8}$$

L Assuming that the LLN holds in the system and using discrete loop equations, find explicitly the limit of the empirical measures $\frac{1}{N} \sum_{i=1}^N \delta_{x_i/N}$ as $M, N \to \infty$ in such a way that $\frac{M}{N} \to \mathbf{m} > 1$.

21.3 Gaussian Fluctuations through Loop Equations

Our next step is to improve the arguments of the previous section, by looking at the lower-order terms of $R_N(Nz)$. Eventually, this provides access to the global fluctuations in the system.

We only sketch the approach; for the detailed exposition, see Borodin et al. (2017). We have

$$\prod_{i=1}^N \left(1 - \frac{1}{N(z - \frac{x_i}{N})} \right) = \exp\left(\sum_{i=1}^N \ln\left(1 - \frac{1}{N(z - \frac{x_i}{N})} \right) \right)$$

$$= \exp\left(-G(z) - \Delta G_N(z) + \frac{1}{2N^2} \sum_{i=1}^N \frac{1}{(z - \frac{x_i}{N})^2} + O(1/N^2) \right),$$

where $\Delta G_N(z)$ is a random fluctuation of the (empirical) Stieltjes transform:

$$\Delta G_N(z) = \frac{1}{N} \sum_{i=1}^N \frac{1}{z - \frac{x_i}{N}} - \int \frac{1}{z - y} \mu(dy).$$

Note that as $N \to \infty$

$$\frac{1}{2N^2} \sum_{i=1}^N \frac{1}{(z - \frac{x_i}{N})^2} = -\frac{1}{2N^2} \frac{\partial}{\partial z} \left(\sum_{i=1}^N \frac{1}{z - \frac{x_i}{N}} \right) = -\frac{1}{2N} G'(z) + o(1).$$

---

[2] This distribution appears on sections of the uniformly random domino tilings of the Aztec diamond, as in Figure 1.10.

Hence,

$$\prod_{i=1}^{N} \left(1 - \frac{1}{N(z - \frac{x_i}{N})}\right) = \exp\left(-G(z) - \Delta G_N(z) - \frac{1}{2N}G'(z) + O(1/N^2)\right).$$

Making a similar computation for the second term of $R_N(Nz)$, we eventually get

$$\frac{R_N(Nz)}{N^2} = \varphi_N^-(Nz)\exp(-G(z)) + \varphi_N^+(Nz)\exp(G(z))$$

$$+ \mathbb{E}[\Delta G_N(z)]\left(-\varphi_N^-(Nz)\exp(-G(z)) + \varphi_N^+(Nz)\exp(G(z))\right)$$

$$+ \frac{1}{N}(\text{some explicit function of } G(z)) + o(1/N), \qquad (21.9)$$

where

$$\varphi_N^+(Nz) = \frac{\varphi^+(Nz)}{N^2} = \left(z + \frac{A-N}{N}\right)\left(\frac{A+N}{N} - z\right),$$

$$\varphi_N^-(Nz) = \frac{\varphi^-(Nz)}{N^2} = z\left(\frac{2A}{N} - z\right).$$

Remark 21.8 We use the lower subscript N in $\varphi_N^\pm(Nz)$ to emphasize that these functions might depend on N for general discrete log-gases. Note that in the hexagon case with scaling $A = \tilde{a}N$, the functions $\varphi_N^\pm(Nz)$ actually are N-independent.

Remark 21.9 The identity (21.9) – or more precisely, the $o(1/N)$ part of it – requires an asymptotic estimate of $\mathbb{E}[\Delta G_N(z)]$. It turns out that $\mathbb{E}[\Delta G_N(z)] = O(1/N)$, but the proof of this fact is a nontrivial argument based on "self-improving" estimates; see Borodin et al. (2017) for details.

The identity (21.9) has the form

$$\frac{R_N(Nz)}{N^2} = R_\infty(z) + \mathbb{E}[\Delta G_N(z)]Q_\infty(z) + \mathcal{E} + \mathcal{S}, \qquad (21.10)$$

where $\mathcal{E}$ is an explicit expression in terms of the limiting $G(z)$, $\mathcal{S}$ is a small term of order $o(1/N)$, and

$$Q_\infty(z) = -\varphi_N^-(Nz)\exp(-G(z)) + \varphi_N^+(Nz)\exp(G(z)).$$

Our next step is to find $\mathbb{E}[\Delta G_N(z)]$ by *solving* (21.9) asymptotically as $N \to \infty$. Let us emphasize that we have no explicit expression for $R_N(Nz)$, but nevertheless, the equation can still be solved by using a bit of complex analysis.

Note that

$$R_\infty(z)^2 - Q_\infty(z)^2 = 4\varphi_N^-(Nz)\varphi_N^+(Nz).$$

In the hexagon case, the RHS of the last identity is a degree four polynomial; hence, $Q_\infty(z)^2$ is a polynomial. But $Q_\infty(z)$ grows linearly as $z \to \infty$. Hence,

$$Q_\infty(z) = \text{const} \cdot \sqrt{(z-p)(z-q)}.$$

One can show, similarly to Exercise 21.6, that p, q in this equation are the intersections of the vertical segment with the boundary of the liquid region.

Take a contour C encircling $[0; 1 + \bar{a}]$, and let w be a point outside of C. Integrating (21.10), we have

$$\frac{\text{const}}{2\pi \mathbf{i}} \oint_C \mathbb{E}[\Delta G_N(z)] \frac{\sqrt{(z-p)(z-q)}}{z-w} dz$$

$$= \frac{1}{2\pi \mathbf{i}} \oint_C \frac{\frac{1}{N^2} R_N(Nz) - R_\infty(z) - \mathcal{E} - \mathcal{S}}{z-w} dz. \quad (21.11)$$

Note that $(\frac{1}{N^2} R_N(Nz) - R_\infty(z))/(z-w)$ is analytic inside C; hence, its integral vanishes (removing the unknown and asymptotically dominating part of the equation). The function

$$\frac{\mathbb{E}[\Delta G_N(z)]\sqrt{(z-p)(z-q)}}{z-w}$$

decays as $O(z^{-2})$ when $z \to \infty$; hence, it has no poles *outside* of C. Therefore, the LHS of (21.11) is equal to minus the residue at w, so that (21.11) transforms into

$$\mathbb{E}[\Delta G_N(w)]\sqrt{(w-p)(w-q)} = (\text{explicit}) + (\text{small}),$$

giving an explicit expression for the first moment of $\Delta G_N(w)$ as $N \to \infty$.

Higher moments of ΔG_N can be computed by repeating this argument for the deformed distribution (21.1) with weight

$$\tilde{w}(x) = w(x) \prod_{a=1}^{k} \left(1 + \frac{t_a}{v_a - \frac{x}{N}}\right),$$

where t_a represents parameters and v_a represents complex numbers. Then, using the previous computation, one can evaluate $\mathbb{E}_{\tilde{\mathbb{P}}}[\Delta G_N]$ for the deformed measure $\tilde{\mathbb{P}}$. On the other hand, consider the following exercise:

Exercise 21.10 Show that by differentiating $\mathbb{E}_{\tilde{\mathbb{P}}}[\Delta G_N]$ by $\frac{\partial^k}{\partial t_1 \cdots \partial t_k}$ at $t_1 = \cdots = t_k = 0$, one gets higher-order *mixed cumulants*[3] of ΔG_N with respect to the original measure.

---

[3] See Definition 23.15 and Exercise 23.16 for some further details.

Further details on how this computation leads to the proof of asymptotic Gaussianity of fluctuations ΔG_N (and explicit evaluation of the covariance of $\Delta G_N(z_1)$ and $\Delta G_N(z_2)$) can be found in Borodin et al. (2017).

21.4 Orthogonal Polynomial Ensembles

The distributions on N-particle configurations of the form (21.1) are also known as "orthogonal polynomial ensembles." The following statement explains the name:

Theorem 21.11 *Consider the probability distribution on N-particle configurations $x_1 < x_2 < \cdots < x_N$, $x_i \in \mathbb{Z}$ of the form*

$$\mathbb{P}(x_1, \ldots, x_N) \propto \prod_{1 \le i < j \le N} (x_i - x_j)^2 \prod_{i=1}^{N} w(x_i), \qquad (21.12)$$

where the weight $w(x)$, $x \in \mathbb{Z}$, either has finite support or decays faster than x^{-2N} as $x \to \infty$. Let $P_k(x)$ be monic orthogonal polynomials with respect to $w(x)$:

$$P_k(x) = x^k + \cdots, \qquad \sum_{x \in \mathbb{Z}} P_k(x) P_\ell(x) w(x) = \delta_{k=\ell} \cdot h_k. \qquad (21.13)$$

Define the Christoffel–Darboux kernel:

$$K(x, y) = \sqrt{w(x)w(y)} \sum_{k=0}^{N-1} \frac{P_k(x) P_k(y)}{h_k}.$$

Then random N particle configuration $\mathcal{X}$ with distribution (21.12) is a determinantal point process with kernel K, which means that for any $m = 1, 2, \ldots,$ and any distinct integers $u_1, \ldots, u_m$

$$\text{Prob}(u_1 \in \mathcal{X}, u_2 \in \mathcal{X}, \ldots, u_m \in \mathcal{X}) = \det\left[K(u_i, u_j)\right]_{i,j=1}^{m}. \qquad (21.14)$$

Sketch of the Proof First, note that the matrix $K(x, y)$ has a rank of N because it can be represented as AA^t for $\mathbb{Z} \times \{0, 1, \ldots, N-1\}$ matrix A with elements $A(x, k) = \sqrt{\frac{w(x)}{h_k}} P_k(x)$. Hence, the RHS of (21.14) vanishes for $m > N$. So does the LHS.

For the case $m = N$, the identity (21.14) relies on the following observation:

Exercise 21.12 Let $P_k(x) = x^k + \cdots$ be any sequence of monic polynomials. Then

$$\det\left[P_{i-1}(x_j)\right]_{i,j=1}^{N} = \prod_{1 \le i < j \le N} (x_j - x_i). \qquad (21.15)$$

Using (21.15) we rewrite (21.12) as

$$\det\left[\frac{\sqrt{w(x_i)}}{\sqrt{h_{a-1}}}P_{a-1}(x_i)\right]_{i,a=1}^{N}\det\left[\frac{\sqrt{w(x_j)}}{\sqrt{h_{a-1}}}P_{a-1}(x_j)\right]_{a,j=1}^{N}. \tag{21.16}$$

Multiplying the matrices under determinants (summing over the a-indices), we get (21.14) up to a constant normalization prefactor, which is implicit in (21.12). For the prefactor verification, we need to check that the sum of (21.16) over arbitrary choices of $x_1, x_2, \ldots, x_N \in \mathbb{Z}$ (note that we do not impose any ordering here) is $N!$. This is done by expanding each determinant as a sum of $N!$ terms, multiplying the two sums, and then summing each product over x_i using (21.13).

For the case $m = N - 1$, using the already-proven $m = N$ case, we need to compute the following sum (in which we set $u_i = x_i$, $i = 1, \ldots, N - 1$, to match (21.14)):

$$\sum_{x_N \in \mathbb{Z}} \det[K(x_i, x_j)]_{i,j=1}^{N}. \tag{21.17}$$

The definition of the kernel $K(x, y)$ implies the following reproducing properties:[4]

$$\sum_{x \in \mathbb{Z}} K(x, x) = N, \qquad \sum_{x \in \mathbb{Z}} K(y, x)K(x, z) = K(y, z). \tag{21.18}$$

Expanding the determinant in (21.17) over the last row and column and using (21.18) to compute the sum over x_N, we get

$$N \det[K(x_i, x_j)]_{i,j=1}^{N-1} + \sum_{a=1}^{N-1}\sum_{b=1}^{N-1}(-1)^{a+b}K(x_a, x_B)\det[K(x_i, x_j)]_{i\neq a; j\neq b}.$$

The double sum is (minus) the sum of $N - 1$ expansions of $\det[K(x_i, x_j)]_{i,j=1}^{N-1}$ over $N - 1$ possible columns. Therefore, (21.17) equals the desired $\det[K(x_i, x_j)]_{i,j=1}^{N-1}$.

The general $1 < m < N$ case is similarly obtained by exploiting the orthogonality relations between polynomials, and we leave it to the reader as an exercise. The details can be found, for example, in Section 5.8 in Mehta (2004), Section 5.4 in Deift (2000), and Section 3 in Borodin and Gorin (2016). $\quad\square$

---

[4] From the functional analysis point of view, these two properties arise because the operator with matrix $K(x, y)$ is an orthogonal projector on the N-dimensional space spanned by the first N orthogonal polynomials. Hence, the trace of this operator is N, and its square is equal to itself.

Let us remark that the particular choice of the lattice $\mathbb{Z}$ to which the points x_i belong in Theorem 21.11 is not important, and the statement remains valid for the orthogonal polynomials on any set. On the other hand, the fact that the interaction is $\prod_{i<j} |x_i - x_j|^2$ rather than $\prod_{i<j} |x_i - x_j|^\beta$ with general $\beta > 0$ is crucial for the validity of this theorem.[5]

Theorem 21.11 reduces the asymptotic questions for $\beta = 2$ log-gases to those for the orthogonal polynomials. Given the importance of this idea, the distribution (19.1) appearing in lozenge tilings of the hexagons is often called the "Hahn ensemble," the distribution (21.8) appearing in domino tilings is called the "Krawtchouk ensemble," and the distribution (19.5) appearing in random Gaussian Hermitian matrices is called the "Hermite ensemble." Each name is derived from the name of the corresponding orthogonal polynomials, and in each case, the polynomials are quite special: they belong to the families of *hypergeometric* orthogonal polynomials, which can be expressed through hypergeometric functions and possess various beneficial properties; see Koekoek et al. (2010) for the Askey scheme of such polynomials and their q-analogues. All the polynomials in the Askey scheme are related by various degenerations, which are reflected in the limit relations between orthogonal polynomial ensembles. For instance, by taking a limit from the Hahn ensemble, one can get the Krawtchouk ensemble, and the latter can be further degenerated into the Hermite ensemble (in Proposition 19.8, we observed a direct limit from the Hahn to the Hermite ensemble). Borodin et al. (2010) found that the q-Racah ensemble (corresponding orthogonal polynomials sit on the very top of the Askey scheme and are related to the basic hypergeometic function $_4\phi_3$) can still be found in the random tilings of a hexagon in the situation when we deal with more complicated measures than the uniform one.

The connections to classical orthogonal polynomials were exploited in Johansson (2002), Baik et al. (2003), Gorin (2008), Borodin et al. (2010), and Breuer and Duits (2017) and applied to study various features of the random lozenge tilings of hexagons. The periodically weighted lozenge tilings (which we briefly mentioned at the end of Lecture 13) require much more sophisticated orthogonal polynomials, and this case is not so well developed; see Charlier et al. (2020) and references therein.

For general reviews of the appearances of orthogonal polynomial ensembles in probability, we refer the reader to Deift (2000), König (2005), and Baik et al. (2016).

---

[5] Similar, yet slightly more complicated, formulas also exist for $\beta = 1, 4$; see Section 5.7 in Mehta (2004).

22

Lecture 22: Plane Partitions and Schur Functions

Lectures 22 and 23 of the class were given by Andrew Ahn.

At the end of the previous lecture we sketched an approach to the analysis of the global fluctuations of random tilings based on the discrete loop equations. When supplied with full details, the approach gives the convergence of fluctuations of the height function of random tilings along a vertical section in the hexagon toward a one-dimensional section of the Gaussian free field.

The aim of this lecture and the next is to present another approach in a slightly different setting of q^{Volume}-weighted plane partitions, which can be viewed as random tilings of certain infinite domains. We are going to prove the full convergence of the centered height function to the Gaussian free field in this setting, thus checking one particular case of Conjecture 11.1.

22.1 Plane Partitions

Fix a $B \times C$ rectangular table. A *plane partition* in the $B \times C$ rectangle is a filling of the table with nonnegative integers, such that the numbers are weakly decreasing along the rows and columns, as in Figure 22.1.

Observe that a plane partition may be viewed as a stepped surface consisting of unit cubes or as a tiling model with three types of lozenges formed by unions of adjacent equilateral triangles. The surface is given by stacking cubes on a plane to a height given by the entries in our $B \times C$ table. The lozenge tiling is given by projecting such an image (diagonally) onto a plane, as in Figure 22.1. If we know a priori that all the numbers in the plane partition are less than or equal to A, then the resulting tiling is nontrivial only inside an $A \times B \times C$ hexagon. However, we are more interested in the case of unbounded entries, in which case the tiling might extend far up, to arbitrary distances and in a

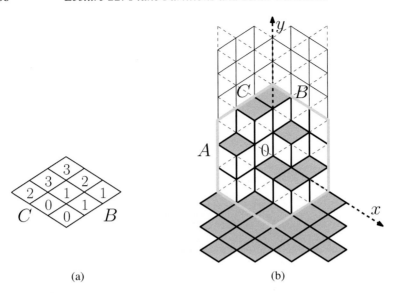

Figure 22.1   Panel (a): A plane partition in a 3×3 rectangle. Panel (b): Corresponding stepped surface, projected onto a two-dimensional plane.

nontrivial way. In any case, far down, the tiling consists solely of horizontal $\diamondsuit$ lozenges, in the far upper right, one sees only $\lozenge$ lozenges, and in the far upper left, one sees only $\lozenge$ lozenges.

From now on, in order to simplify the exposition, we consider only the symmetric case $B = C$ and deal with plane partitions in an $N \times N$ square (the same methods work for $B \neq C$ as well). On the plane, we use the coordinate system shown in panel (b) of Figure 22.1, with $(0, 0)$ placed at a distance N in the positive vertical direction from the bottom tip of the tiling.

We define a modified height function $h \colon \mathbb{Z} \times \mathbb{Z} \to \mathbb{Z}_{\geq 0}$. We set $h(x, -\infty) = 0$ and let $h(x, y) - h(x, y - 1) = 0$ whenever the vector $(x, y - 1) - (x, y)$ crosses a horizontal lozenge $\diamondsuit$ and $h(x, y) - h(x, y - 1) = 1$ if the vector $(x, y - 1) - (x, y)$ follows an edge of $\lozenge$ or $\lozenge$ lozenges. In other words, $h(x, y)$ counts the total number of nonintersecting paths (following $\lozenge$, $\lozenge$ lozenges; see Figures 2.2 and 18.1) below (x, y). This height function differs from the one of Lecture 1 by an affine transformation.

Exercise 22.1   Find explicitly the transformation that matches the modified height function of this section with the one given by (1.3) in Lecture 1.

Unlike the tilings of a given hexagon, the total number of plane partitions in an $N \times N$ square is infinite. Thus, there is no uniform probability measure on tilings. Instead, we study the following q-deformed version:

Definition 22.2 Let Volume(π) be the volume of a plane partition π, which is the sum of all N^2 integers in the table. Given $0 < q < 1$, define the q^{Volume} probability measure on a plane partition in the $N \times N$ square through

$$\mathbb{P}_{N,q}(\pi) \propto q^{\text{Volume}(\pi)}.$$

If the parameter q is small, then a typical $\mathbb{P}_{N,q}$ random partition has just a few boxes. On the other hand, as $q \to 1$, the number of boxes grows to infinity, and we can expect nontrivial asymptotic theory.

22.2 Schur Polynomials

Our treatment of random plane partitions uses Schur polynomials and their combinatorics. We refer the reader to Chapter I in Macdonald (1995) for the detailed treatment of symmetric polynomials and only review the necessary facts here.

Let $\mathbb{GT}_N$ denote the set of all N-tuples of integers $\lambda_1 \geq \lambda_2 \geq \cdots \geq \lambda_N$. In representation-theoretic context, elements of $\mathbb{GT}_N$ are often called "signatures," and the notation $\mathbb{GT}$ comes from the names of Gelfand and Tsetlin, who introduced an important basis (parameterized by a sequence of signatures) in an irreducible representation of the unitary group $U(N)$.

Definition 22.3 Given $\vec{k} = (k_1, \ldots, k_N) \in \mathbb{Z}^N$, define

$$a_{\vec{k}}(x_1, \ldots, x_N) = \det \begin{pmatrix} x_1^{k_1} & \cdots & x_1^{k_N} \\ \vdots & \ddots & \vdots \\ x_N^{k_1} & \cdots & x_N^{k_N} \end{pmatrix} = \sum_{\sigma \in S_N} \text{sign}(\sigma) x_1^{k_{\sigma(1)}} \cdots x_N^{k_{\sigma(N)}},$$

where the sum is taken over all permutations. Given an $\lambda \in \mathbb{GT}_N$ and setting $\delta = (N-1, N-2, \ldots, 0)$, define the *Schur* symmetric (Laurent) polynomials through

$$s_\lambda(x_1, \ldots, x_N) = \frac{a_{\lambda+\delta}(x_1, \ldots, x_N)}{a_\delta(x_1, \ldots, x_N)}.$$

Exercise 22.4 The denominator a_δ is known as the "Vandermonde determinant." Check that

$$a_\delta(x_1, \ldots, x_N) = \prod_{1 \leq i < j \leq N} (x_i - x_j).$$

In order to see that $s_\lambda(x_1, \ldots, x_N)$ is, indeed, a (Laurent) polynomial, notice that $a_{\vec{k}}$ vanishes whenever $x_i = x_j$; hence, it must be divisible by each factor $(x_i - x_j)$, $1 \leq i < j \leq N$, and, therefore, also by their product.

Two signatures $\lambda \in \mathbb{GT}_N$ and $\mu \in \mathbb{GT}_{N-1}$ *interlace*, denoted $\mu \prec \lambda$, if

$$\lambda_1 \geq \mu_1 \geq \lambda_2 \geq \mu_2 \geq \cdots \geq \mu_{N-1} \geq \lambda_N.$$

Lemma 22.5 (Branching Rule)   *For any $\lambda \in \mathbb{GT}_N$, we have*

$$s_\lambda(x_1, \ldots, x_N) = \sum_{\mu \in \mathbb{GT}_{N-1}, \, \mu \prec \lambda} x_1^{|\lambda| - |\mu|} s_\mu(x_2, \ldots, x_N), \tag{22.1}$$

where $|\lambda| = \lambda_1 + \cdots + \lambda_N$ *and* $|\mu| = \mu_1 + \cdots + \mu_N$.

Proof   If $\vec{\ell} = (\lambda_1 + N - 1, \ldots, \lambda_N)$ and $\vec{m} = (\mu_1 + N - 2, \ldots, \mu_{N-1})$, then $\mu \prec \lambda$ means that

$$\ell_1 > m_1 \geq \ell_2 > m_2 \geq \cdots \geq \ell_N.$$

Multiplying the right-hand side of (22.1) by $a_\delta(x_1, \ldots, x_N)$ gives

$$\sum_{m_1=\ell_2}^{\ell_1-1} \cdots \sum_{m_{N-1}=\ell_N}^{\ell_{N-1}-1} x_1^{|\lambda|-|\mu|} \sum_{\tau \in S_{N-1}} \mathbf{sign}(\tau) x_{\tau(2)}^{m_1} \cdots x_{\tau(N)}^{m_{N-1}} \prod_{i=2}^{N} (x_1 - x_i),$$

where we consider the action of $\tau \in S_{N-1}$ on $\{2, \ldots, N\}$, so that $2 \leq \tau(i) \leq N$ for $i = 2, 3, \ldots, N$. Writing

$$|\lambda| - |\mu| = (\ell_1 - m_1 - 1) + \cdots + (\ell_{N-1} - m_{N-1} - 1) + \ell_N,$$

the summation becomes

$$\sum_{\tau \in S_{N-1}} \mathbf{sign}(\tau) x_1^{\ell_N} \sum_{m_1=\ell_2}^{\ell_1-1} x_1^{\ell_1-m_1-1} x_{\tau(2)}^{m_1} (x_1 - x_{\tau(2)})$$

$$\cdots \sum_{m_{N-1}=\ell_N}^{\ell_{N-1}} x_1^{\ell_{N-1}-m_{N-1}-1} x_{\tau(N)}^{m_N} (x_1 - x_{\tau(N)}).$$

Each interior summation telescopes, so we obtain

$$\sum_{\tau \in S_{N-1}} \mathbf{sign}(\tau) x_1^{\ell_N} (x_1^{\ell_1-\ell_2} x_{\tau(2)}^{\ell_2} - x_{\tau(2)}^{\ell_1}) \cdots (x_1^{\ell_{N-1}-\ell_N} x_{\tau(N)}^{\ell_N} - x_{\tau(N)}^{\ell_{N-1}}).$$

The last sum is the determinant of the following $N \times N$ matrix:

$$\begin{pmatrix} 0 & x_2^{\ell_1} - x_1^{\ell_1-\ell_2} x_2^{\ell_2} & \cdots & x_N^{\ell_1} - x_1^{\ell_1-\ell_2} x_N^{\ell_2} \\ & & \vdots & \\ 0 & x_2^{\ell_{N-1}} - x_1^{\ell_{N-1}-\ell_N} x_2^{\ell_N} & \cdots & x_N^{\ell_{N-1}} - x_1^{\ell_{N-1}-\ell_N} x_N^{\ell_N} \\ x_1^{\ell_N} & x_2^{\ell_N} & \cdots & x_N^{\ell_N} \end{pmatrix}.$$

We perform elementary manipulations preserving the determinant on the last matrix. Adding to the $(N - 1)$st row the Nth row multiplied by $x_1^{\ell_{N-1}-\ell_N}$, we bring the $(N-1)$st row to the form $x_j^{\ell_{N-1}}$. Further adding to $(N-2)$st row a proper multiple of the $(N-1)$st row, we bring the $(N-2)$st row to the form $x_j^{\ell_{N-2}}$. Further repeating this procedure, we bring the matrix to the form $(x_j^{\ell_i})_{i,j=1}^N$, matching the left-hand side of (22.1). □

Lemma 22.6 (Cauchy Identity) *For $|x_1|,\ldots,|x_N|,|y_1|,\ldots,|y_N| < 1$, we have*

$$\prod_{i,j=1}^N \frac{1}{1-x_iy_j} = \sum_{\lambda\in\mathrm{GT}_N,\,\lambda_N\geq 0} s_\lambda(x_1,\ldots,x_N)s_\lambda(y_1,\ldots,y_N).$$

Proof The Cauchy determinant formula of (20.7) reads as follows:

$$\det\left[\frac{1}{u_i-v_j}\right]_{i,j=1}^N = \frac{\prod_{1\leq i<j\leq N}(u_i-u_j)(v_j-v_i)}{\prod_{i,j=1}^N(u_i-v_j)},$$

which implies

$$\det\left[\frac{1}{1-x_iy_j}\right]_{i,j=1}^N = \frac{\prod_{1\leq i<j\leq N}(x_i-x_j)(y_i-y_j)}{\prod_{i,j=1}^N(1-x_iy_j)} = a_\delta(\vec{x})a_\delta(\vec{y})\prod_{i,j=1}^N(1-x_iy_j)^{-1}.$$

On the other hand,

$$\det\left[\frac{1}{1-x_iy_j}\right]_{i,j=1}^N = \det\left[\sum_{n=0}^\infty(x_iy_j)^n\right]_{i,j=1}^N = \sum_{\alpha_1=0}^\infty\cdots\sum_{\alpha_N=0}^\infty \det\left[x_i^{\alpha_j}y_j^{\alpha_j}\right]_{i,j=1}^N$$

$$= \sum_{\alpha_1=0}^\infty\cdots\sum_{\alpha_N=0}^\infty a_\alpha(\vec{x})y_1^{\alpha_1}\cdots y_N^{\alpha_N} = \sum_{\lambda\in\mathrm{GT}_N,\,\lambda_N\geq 0} a_{\lambda+\delta}(\vec{x})a_{\lambda+\delta}(\vec{y}). \quad (22.2)$$

Dividing by $a_\delta(\vec{x})$, $a_\delta(\vec{y})$ completes the proof. □

22.3 Expectations of Observables

We do not have direct access to the distribution of the value of the height function of random plane partitions at a particular point. However, we can produce contour-integral formulas for the expectations of certain polynomial sums in these values, which we call "observables." They are rich enough to fully characterize the distribution and therefore are useful for the asymptotic analysis.

The main purpose of this section is to obtain the following formula for the expectations of observables of the random height function distributed according to $\mathbb{P}_{N,q}$:

Theorem 22.7 *Let $-N < x_1 \le \cdots \le x_k < N$ and $0 < t_1, \ldots, t_k < 1$. Then*

$$\mathbb{E}\left[\prod_{i=1}^{k}\sum_{y\in\mathbb{Z}}h(x_i,y)t_i^y\right] = \frac{1}{(2\pi\mathbf{i})^k}\oint\cdots\oint\prod_{1\le i<j\le k}\frac{(z_j-\frac{t_i}{t_j}z_i)(z_j-z_i)}{(z_j-t_iz_i)(z_j-\frac{1}{t_j}z_i)}$$

$$\times\prod_{i=1}^{k}G_{x_i}(z_i;t_i)\frac{dz_i}{(t_i^{-1}-1)(1-t_i)z_i},\qquad(22.3)$$

where

$$G_x(z;t)=t^{\min(0,x)}\prod_{\substack{i\in\mathbb{Z}+\frac{1}{2}\\-N<i<\min(0,x)}}\frac{1-t^{-1}q^{-i}z^{-1}}{1-q^{-i}z^{-1}}\prod_{\substack{i\in\mathbb{Z}+\frac{1}{2}\\\max(0,x)<i<N}}\frac{1-q^iz}{1-tq^iz}.$$

Note that the sum $\sum_{y\in\mathbb{Z}}h(x_i,y)t_i^y$ is infinite only in the positive y direction because $h(x_i,y)$ vanishes for sufficiently small y. The contours satisfy the following properties:

- *The z_i-contour includes the poles at $0, q^{N-\frac{1}{2}}, q^{N-\frac{3}{2}}, \ldots, q^{-\min(0,x_i)-\frac{1}{2}}$ but does not contain any of the poles at $t_i^{-1}q^{-\max(0,x_i)-\frac{1}{2}}, t_i^{-1}q^{-\max(0,x_i)-\frac{3}{2}}, \ldots, t_i^{-1}q^{-N+\frac{1}{2}}$.*
- *For $1 \le i < j \le k$, we have $|z_i| < |t_jz_j|$ (implying also $|t_iz_i| < |z_j|$) everywhere on the contours.*

Exercise 22.8 Assume $x_1 = x_2 = \cdots = x_k$. Show that the residue of the integrand in (22.3) at $z_i = t_jz_j$, $1 \le i < j \le k$ (but not for $i > j$) vanishes. In other words, it does not matter whether the integration contours are nested in such a way that $|z_i| < |t_jz_j|$ or $|z_i| > |t_jz_j|$, $1 \le i < j \le k$; the integral is the same for both choices. (You should still have $|t_iz_i| < |z_j|$ though.)

Remark 22.9 A *skew* plane partition is a generalization in which we place integers not in a rectangular table but in a difference between a rectangle and a Young diagram. Theorem 22.7 has an extension to q^{Volume}-weighted skew plane partitions; see Ahn (2018) for the details.

In the rest of the lecture we outline the main ideas for the proof of the theorem.

We first observe that a plane partition π in the $N \times N$ square grid may be viewed as a sequence of interlacing signatures:

$$(\emptyset) = \pi^{-N} \prec \pi^{-N+1} \prec \cdots \prec \pi^{-1} \prec \pi^0 \succ \pi^1 \succ \cdots \succ \pi^{N-1} \succ \pi^N = (\emptyset).$$
$$(22.4)$$

In the notations in panel (a) of Figure 22.1, each signature represents a vertical slice of the picture, so that $\pi^{-N+1} \in GT_1, \ldots, \pi^0 \in GT_N, \ldots, \pi^{N-1} \in GT_1$. The slices are read from left to right. In particular, the plane partition of Figure 22.1 corresponds to the sequence

$$(\emptyset) \prec (2) \prec (3, 0) \prec (3, 1, 0) \succ (2, 1) \succ (1) \succ (\emptyset).$$

In what follows, we often treat a signature $\lambda \in GT_N$ (with nonnegative coordinates) as a signature $\mu \in GT_M$ with $M > N$ by adding sufficiently many zero last coordinates. For instance, $(2, 1) \in GT_2$ can be treated as $(2, 1, 0) \in GT_3$ or as $(2, 1, 0, 0) \in GT_4$.

Lemma 22.10 *Fix $x \in \{-N + 1, -N + 2, \ldots, N - 1\}$ and take arbitrary $M \geq N - |x|$. Let $\lambda \in GT_M$ be the signature encoding the vertical slice of the plane partition at abscissa x (i.e., λ is π^x with $M - N + |x|$ additional zeros). For any $0 < t < 1$, we have*

$$\sum_{y \in \mathbb{Z}} h(x, y) t^y = \frac{t^{\min(0, x)}}{1 - t} \left(\sum_{i=1}^{M} t^{\lambda_i - i + 1} - \frac{t^{-M+1}}{1 - t} \right).$$

Proof The appearance of $\min(0, x)$ in the formulas is related to our particular choice of the coordinate system, as in panel (b) of Figure 22.1. The ith from the top horizontal lozenge in the vertical section of the tilings at abscissa x has top and bottom coordinates $\pi_i^x - i + 1 + \min(0, x)$ and $\pi_i^x - i + \min(0, x)$, respectively. With this in mind and using that $h(x, y) - h(x, y - 1) = 1$ unless y corresponds to the top of a horizontal lozenge, we compute

$$\sum_{y \in \mathbb{Z}} h(x, y) t^y = \frac{1}{1 - t} \sum_{y \in \mathbb{Z}} h(x, y)(t^y - t^{y+1})$$

$$= \frac{1}{1 - t} \sum_{y \in \mathbb{Z}} (h(x, y) - h(x, y - 1)) t^y$$

$$= \frac{1}{1 - t} \sum_{y \in \mathbb{Z}} (1 - \mathbb{1}[y \in \{\pi_i^x - i + 1 + \min(0, x)\}_{i=1}^{\infty}]) t^y$$

$$= \frac{1}{1 - t} \sum_{y > -M + \min(0, x)} (1 - \mathbb{1}[y \in \{\pi_i^x - i + 1 + \min(0, x)\}_{i=1}^{M}]) t^y$$

$$= \frac{1}{1 - t} \left(-\sum_{i=1}^{M} t^{\pi_i^x - i + 1 + \min(0, x)} + \frac{t^{-M+1+\min(0, x)}}{1 - t} \right),$$

where in the third line, π^x is extended to have infinitely many coordinates by adding (infinitely many) zeros; in the fourth line, π^x is extended to an element of $\mathbb{GT}_M$ (by adding $M - N + |x|$ zeros); and the last line is obtained from the fourth by summing ones and indicators $\mathbb{1}$ separately. □

In view of the previous lemma, set

$$\wp_t(\lambda) = \sum_{i=1}^{M} t^{\lambda_i - i + 1} - \frac{t^{-M+1}}{1-t}, \qquad \lambda \in \mathbb{GT}_M, \lambda_M \geq 0.$$

Note that if we increase M and simultaneously extend λ by adding zeros, then $\wp_t(\lambda)$ is unchanged.

Our plan is to compute expectations of products of $\wp_t(\pi^x)$ and then use the previous lemma to obtain Theorem 22.7. This is achieved using the family of difference operators described in the definition that follows.

Definition 22.11   The first *Macdonald* operator $D_t^{\{x_1,\ldots,x_n\}}$ is defined by

$$D_t^{\{x_1,\ldots,x_n\}} = \sum_{i=1}^{n} \left[\prod_{j \neq i} \frac{x_i - t^{-1}x_j}{x_i - x_j} \right] T_{t,x_i} - \frac{t^{-n+1}}{1-t},$$

where T_{t,x_i} is the t-shift operator, which maps

$$F(x_1,\ldots,x_{i-1},x_i,x_{i+1},\ldots,x_N) \mapsto F(x_1,\ldots,x_{i-1},tx_i,x_{i+1},\ldots,x_N).$$

Remark 22.12   The operators $D_t^{\{x_1,\ldots,x_n\}}$ are $q = t$ versions of the difference operators used by Ian Macdonald in his study of the (q,t)-deformation of Schur polynomials bearing his name; see Chapter VI in Macdonald (1995).

Lemma 22.13   *The operator $D_t^{\{x_1,\ldots,x_n\}}$ is diagonalized by the Schur functions $s_\lambda(x_1,\ldots,x_n)$, $\lambda \in \mathbb{GT}_n$. We have*

$$D_t^{\{x_1,\ldots,x_n\}} s_\lambda(x_1,\ldots,x_n) = \wp_t(\lambda) s_\lambda(x_1,\ldots,x_n).$$

Proof   Observe that

$$\left(D_t^{\vec{x}} + \frac{t^{-n+1}}{1-t} \right) s_\lambda(\vec{x}) = t^{-n+1} \sum_{i=1}^{n} \left[\prod_{j \neq i} \frac{tx_i - x_j}{x_i - x_j} \right] \frac{T_{t,x_i} a_{\lambda+\delta}(\vec{x})}{T_{t,x_i} a_\delta(\vec{x})} = \sum_{i=1}^{n} \frac{T_{t,x_i} a_{\lambda+\delta}(\vec{x})}{a_\delta(\vec{x})},$$

where the second equality uses the fact that the $\frac{tx_i - x_j}{x_i - x_j}$ terms replace the t-shifted factors in $T_{t,x_i} a_\delta(\vec{x})$ with the corresponding ordinary factors in $a_\delta(\vec{x})$. Writing the alternant as

$$a_{\lambda+\delta}(\vec{x}) = \sum_{\sigma \in S_n} \operatorname{sgn}(\sigma) \prod_{j=1}^{n} x_j^{\lambda_{\sigma(j)} + n - \sigma(j)},$$

we have that

$$\sum_{i=1}^{n} T_{t,x_i} a_{\lambda+\delta}(\vec{x}) = \left(\sum_{i=1}^{n} t^{\lambda_i+n-i}\right) a_{\lambda+\delta}(\vec{x}),$$

which implies the statement of the lemma. ◻

We are able to produce a contour-integral representation for the action of these operators on *multiplicative* functions $\prod_{i=1}^{n} f(x_i)$, as in the lemma that follows.

Lemma 22.14 (Borodin and Corwin, 2014) *Let $f(z)$ be analytic and $q \in \mathbb{C}$ such that $\frac{f(qz)}{f(z)}$ has no singularities in a neighborhood of $\{0\} \cup \{x_i\}_{i=1}^{n}$. Then*

$$D_t^{\vec{x}} \prod_{i=1}^{n} f(x_i) = \prod_{i=1}^{n} f(x_i) \oint \prod_{j=1}^{n} \frac{z - t^{-1}x_j}{z - x_j} \cdot \frac{f(tz)}{f(z)} \frac{dz}{(1-t^{-1})z},$$

where the contour contains 0 and $\{x_i\}_{i=1}^{n}$ and no other singularities of the integrand.

Proof This follows from the residue theorem and the definition of the operator $D_t^{\vec{x}}$. ◻

The idea of the proof of Theorem 22.7 is to use the D_t operators to extract the volume information, relying on the following lemma:

Lemma 22.15 *Let F denote a function of $2N$ variables $a_1, \ldots, a_N, b_1, \ldots b_N$ given by*

$$F = \prod_{i,j=1}^{N} \frac{1}{1 - a_i b_j}. \tag{22.5}$$

Then, for q^{Volume} random plane partitions identified with interlacing sequences of signatures through (22.4), any $-N < x_1 \leq \cdots \leq x_k \leq 0$, $N > x_1' \geq \cdots \geq x_{k'}' \geq 0$, and any $0 < t_i, t_i' < 1$, we have

$$\mathbb{E}\left[\wp_{t_1}(\pi^{x_1}) \cdots \wp_{t_k}(\pi^{x_k}) \cdot \wp_{t_1'}(\pi^{x_1'}) \cdots \wp_{t_{k'}'}(\pi^{x_{k'}'})\right]$$

$$= \left[F^{-1} D_{t_1}^{\{a_N,\ldots,a_{N+x_1}\}} \cdots D_{t_k}^{\{a_N,\ldots,a_{N+x_k}\}} D_{t_1'}^{\{b_N,\ldots,b_{x_1'-1}\}} \cdots \right. \tag{22.6}$$

$$\left. \cdots D_{t_k'}^{\{b_N,\ldots,b_{x_k'-1}\}} F\right]_{a_i=b_i=q^{i-1/2},\, i=1,\ldots,N}. \tag{22.7}$$

Exercise 22.16 Given Lemma 22.15, prove Theorem 22.7 by a recursive application of Lemma 22.14 to convert the right-hand side of (22.6) into contour integrals.

Let us present a historical overview for Lemma 22.15. The connection between plane partitions and Schur polynomials was first exploited in a probabilistic context in Okounkov and Reshetikhin (2003). The use of the differential operators for the case $k' = 0$ and $x_1 = x_2 = \cdots = x_k$ was introduced in Borodin and Corwin (2014). The extension to arbitrary x_i (still $k' = 0$) was presented in Section 4 of Borodin et al. (2016). An observation that $k' > 0$ is also possible can be found in Section 3 of the same article, which also presents another way to reach Theorem 22.14 that avoids the $k' > 0$ case, and Proposition 2.2.4 in Aggarwal (2015) is a generalization of Theorem 22.14 obtained by this method. We also refer the reader to Ahn (2018), where another set of difference operators is used to extract the observables of plane partitions.

We only present the $k = 1$, $k' = 0$ case of Lemma 22.15; the general case is an extension of the arguments presented here, and details can be found in the aforementioned references.

Proof of $k = 1$, $k' = 0$ Case of Lemma 22.15   The first observation is that with the notation $|\pi^i| = \pi_1^i + \pi_2^i + \cdots$, we have

$$q^{\text{Volume}(\pi)} = q^{\frac{1}{2}(|\pi_0|-|\pi_1|)} q^{\frac{3}{2}(|\pi_1|-|\pi_2|)} \cdots q^{(N-\frac{1}{2})|\pi_{N-1}|}$$
$$\times q^{\frac{1}{2}(|\pi_0|-|\pi_{-1}|)} q^{\frac{3}{2}(|\pi_{-1}|-|\pi_{-2}|)} \cdots q^{(N-\frac{1}{2})|\pi_{-N+1}|}.$$

Hence, if we take the expression

$$\sum_{\lambda \in \text{GT}_N,\, \lambda_N \geq 0} s_\lambda\left(q^{N-\frac{1}{2}}, q^{N-\frac{3}{2}}, \ldots, q^{1/2}\right) s_\lambda\left(q^{N-\frac{1}{2}}, q^{N-\frac{3}{2}}, \ldots, q^{1/2}\right)$$

and expand both s_λ factors into sums of monomials by recursively using the branching rule of Lemma 22.5, then each term in the expanded sum corresponds to a plane partition as in (22.4), and the value of such term is precisely $q^{\text{Volume}(\pi)}$.

We can also first expand

$$\sum_{\lambda \in \text{GT}_N,\, \lambda_N \geq 0} s_\lambda(a_1, \ldots, a_N)\, s_\lambda(b_1, \ldots, b_N) \tag{22.8}$$

into monomials in a_i, b_i, getting a sum over plane partitions, and substitute $a_i = b_i = q^{i-1/2}$ later on. Note that (22.8) evaluates to the function F of (22.5) by Lemma 22.6. In particular, we get

$$[F]_{a_i=b_i=q^{i-1/1}} = \sum_\pi q^{\text{Volume}(\pi)}. \tag{22.9}$$

Assume without loss of generality that $x_1 \leq 0$. Let us expand (22.8) into monomials corresponding to plane partitions π, then fix $\pi^{x_1}, \pi^{x_1+1}, \ldots, \pi^{N-1}$

and make a summation over $\pi^{1-N}, \pi^{2-N}, \ldots, \pi^{x_1-1}$. We claim that the result has the form

$$s_{\pi^{x_1}}(a_N, a_{N-1}, \ldots, a_{1-x_1}) \times \text{(terms depending on } a_{-x_1}, \ldots, a_1, \; b_1, \ldots, b_N). \tag{22.10}$$

Indeed, the summation goes over interlacing signatures; hence, we can again use Lemma 22.5 to turn the result into the Schur function. We can now act with $D_t^{\{a_N, a_{N-1}, \ldots, a_{1-x_1}\}}$ on F by using (22.10) and the eigenrelation of Lemma 22.13. The factor $\wp_t(\pi^{x_1})$ pops out, and, afterward, we can again expand $s_{\pi^{x_1}}(a_N, a_{N-1}, \ldots, a_{1-x_1})$ into monomials. The conclusion is that

$$\left[D_t^{\{a_N, a_{N-1}, \ldots, a_{1-x_1}\}} F \right]_{a_i = b_i = q^{i-1/2}} = \sum_{\pi} \wp_t(\pi^{x_1}) q^{\text{Volume}(\pi)}.$$

Dividing by F and using (22.9), we get the desired evaluation of $\mathbb{E}\, \wp_t(\pi^{x_1})$. □

Exercise 22.17 Using (22.9), compute the partition function of q^{Volume}-weighted plane partitions; that is, show that[1]

$$\sum_{\text{plane partitions}} q^{\text{Volume}} = \prod_{n=1}^{\infty} (1 - q^n)^{-n}, \tag{22.11}$$

where the sum is taken over *all* plane partitions, which is the $N = \infty$ version of the ones in Definition 22.2.

---

[1] Similarly to (1.1) this formula also bears the name of MacMahon and has been known for more than 100 years, although the first proofs were not using Schur polynomials.

23

Lecture 23: Limit Shape and Fluctuations for Plane Partitions

In this lecture we use Theorem 22.7 to analyze global asymptotics (law of large numbers [LLN] and central limit theorem [CLT]) for random q^{Volume}-weighted plane partitions.

In what follows, we fix $\hat{N} > 0$ and set $q = e^{-\varepsilon}$, $N = \lfloor \hat{N}/\varepsilon \rfloor$ for a small parameter $\varepsilon > 0$.

Our goal is to show that for the random height function $h(x, y)$ of plane partitions in an $N \times N$ rectangle and distributed according to $\mathbb{P}_{N,q}$ of Definition 22.2, as $\varepsilon \to 0$, we have

$$\varepsilon h \left(\frac{\hat{x}}{\varepsilon}, \frac{\hat{y}}{\varepsilon} \right) \to h^{\infty}(\hat{x}, \hat{y}),$$

$$h \left(\frac{\hat{x}}{\varepsilon}, \frac{\hat{y}}{\varepsilon} \right) - \mathbb{E}\, h \left(\frac{\hat{x}}{\varepsilon}, \frac{\hat{y}}{\varepsilon} \right) \to \text{GFF},$$

where h^{∞} is a deterministic limit shape, and GFF is the Gaussian free field governing fluctuations, as in Lectures 11 and 12.

23.1 Law of Large Numbers

Our first result is the limit-shape theorem for the height function of the q^{Volume}-weighted random plane partitions in an $N \times N$ rectangle.

Theorem 23.1 *For any $\hat{x} \in [-\hat{N}, \hat{N}]$ and $c > 0$, we have*

$$\lim_{\varepsilon \to 0} \mathbb{E} \left[\sum_{\hat{y} \in \varepsilon \mathbb{Z}} \varepsilon^2 h \left(\frac{\hat{x}}{\varepsilon}, \frac{\hat{y}}{\varepsilon} \right) e^{-c\hat{y}} \right] = \int h^{\infty}(\hat{x}, \hat{y}) e^{-c\hat{y}} \, d\hat{y}, \qquad (23.1)$$

where h^{∞} is an explicit function described in Theorem 23.5.

Remark 23.2   In the next section we show that the variance of the height function decays, which automatically upgrades the convergence of expectations in (23.1) to convergence in probability.

Remark 23.3   Formally, the height function $h(x, y)$ is defined only at integer points (x, y), and therefore, we should write $h\left(\left\lfloor\frac{\hat{x}}{\varepsilon}\right\rfloor, \left\lfloor\frac{\hat{y}}{\varepsilon}\right\rfloor\right)$, where $\lfloor\cdot\rfloor$ is the integer part. However, we are going to omit the integer part in order to simplify the notations.

The proof of Theorem 23.1 starts by expressing (23.1) as a contour integral.

Lemma 23.4   *For any $c > 0$ we have*

$$\lim_{\varepsilon \to 0} \mathbb{E}\left[\sum_{\hat{y} \in \varepsilon \mathbb{Z}} \varepsilon^2 h\left(\frac{\hat{x}}{\varepsilon}, \frac{\hat{y}}{\varepsilon}\right) e^{-cy}\right] = \frac{1}{2\pi i} \oint_0 \mathcal{G}_{\hat{x}}(z)^c \frac{dz}{c^2 z}, \qquad (23.2)$$

where

$$\mathcal{G}_{\hat{x}}(z) = \frac{1 - z}{e^{-\hat{N}} - z} \cdot \frac{e^{-\hat{x}} - z^{-1}}{e^{-\hat{N}} - z^{-1}}, \qquad (23.3)$$

where the branch for raising to the cth power is the one giving positive real values to $z = 0$ and $z = \infty$, and integration goes over a positively oriented contour enclosing 0 and $[e^{-\hat{N}}, e^{\min(0,\hat{x})}]$ but not $[e^{\max(0,\hat{x})}, e^{\hat{N}}]$.

Proof   The $k = 1$ case of Theorem 22.7 reads

$$\mathbb{E}\left[\sum_{y \in \mathbb{Z}} h(x, y)t^y\right] = \frac{1}{2\pi i} \oint G_x(z; i) \frac{dz}{(t^{-1} - 1)(1 - t)z}, \qquad (23.4)$$

where

$$G_x(z; t) = t^{\min(0,x)} \prod_{\substack{i \in \mathbb{Z}+\frac{1}{2} \\ -N < i < \min(0,x)}} \frac{1 - t^{-1}q^{-i}z^{-1}}{1 - q^{-i}z^{-1}} \prod_{\substack{i \in \mathbb{Z}+\frac{1}{2} \\ \max(0,x) < i < N}} \frac{1 - q^i z}{1 - tq^i z},$$

and the integration contour includes the poles at $0, q^{N-\frac{1}{2}}, q^{N-\frac{3}{2}}, \ldots, q^{\min(0,x)-\frac{1}{2}}$.

We set

$$t = \exp(-c\varepsilon), \quad q = \exp(-\varepsilon), \quad x = \frac{\hat{x}}{\varepsilon}, \quad y = \frac{\hat{y}}{\varepsilon}, \quad N = \frac{\hat{N}}{\varepsilon}$$

and send $\varepsilon \to 0$. Let us analyze the asymptotic behavior of $G_x(z; t)$. Clearly, $t^{\min(0,x)} \to e^{-c\min(0,\hat{x})}$. Using the asymptotic expansion $\ln(1 + u) = u + o(u)$, the logarithm of the first product in $G_x(z,t)$ behaves as follows:

$$\sum_{\substack{i\in\mathbb{Z}+\frac{1}{2}\\-\hat{N}/\varepsilon<i<\min(0,\hat{x}/\varepsilon)}} \ln\left(\frac{1-t^{-1}q^{-i}z^{-1}}{1-q^{-i}z^{-1}}\right)$$

$$=\sum_{\substack{i\in\mathbb{Z}+\frac{1}{2}\\-\hat{N}/\varepsilon<i<\min(0,\hat{x}/\varepsilon)}} (1-t^{-1})\frac{q^{-i}z^{-1}}{1-q^{-i}z^{-1}}+o(1)$$

$$=\frac{1-t^{-1}}{1-q}\sum_{\substack{i\in\mathbb{Z}+\frac{1}{2}\\-\hat{N}/\varepsilon<i<\min(0,\hat{x}/\varepsilon)}} (q^{-i}-q^{-i+1})\frac{z^{-1}}{1-q^{-i}z^{-1}}+o(1)$$

$$=-c\int_{e^{-\hat{N}}}^{e^{\min(0,\hat{x})}}\frac{z^{-1}}{1-uz^{-1}}\,du+o(1)=c\int_{e^{-\hat{N}}}^{e^{\min(0,\hat{x})}}\frac{1}{u-z}\,du+o(1)$$

$$=c\ln\left(\frac{e^{\min(0,\hat{x})}-z}{e^{-\hat{N}}-z}\right)+o(1).$$

Note that the logarithm of the complex argument is a mutlivalued function, and here we mean the branch that gives 0 for $z=\infty$ in the last formula (and a real number at $z=0$). The convergence is uniform over z in compact subsets of $\mathbb{C}\setminus[\exp(-\hat{N}),\exp(\min(0,\hat{x}))]$.

Similarly, the asymptotic behavior of the logarithm of the second product in $G_x(z,t)$ is

$$c\ln\left(\frac{e^{-\max(0,\hat{x})}-z^{-1}}{e^{-\hat{N}}-z^{-1}}\right),$$

with uniform convergence on compact subsets of $\mathbb{C}\setminus[e^{\max(0,\hat{x})},e^{\hat{N}}]$ and the branch of the logarithm, which gives 0 at $z=0$ in the last formula (and a real number at $z=\infty$). Combining the factors, we get

$$\lim_{\varepsilon\to0}G_{\tilde{x}/\varepsilon}(z;t)=\left(e^{-\min(0,\hat{x})}\frac{e^{\min(0,\hat{x})}-z}{e^{-\hat{N}}-z}\cdot\frac{e^{-\max(0,\hat{x})}-z^{-1}}{e^{-\hat{N}}-z^{-1}}\right)^c,\qquad(23.5)$$

where the convergence is uniform over compact subsets of $\mathbb{C}\setminus[\exp(-\hat{N}),\exp(\min(0,\hat{x}))]\setminus[e^{\max(0,\hat{x})},e^{\hat{N}}]$, and the branch for the cth power is the one giving positive real values for $z=0$ and $z=\infty$. Checking cases $\hat{x}\le0$ and $\hat{x}\ge0$ separately, one sees that (23.5) is the same expression as

$$\left(\frac{1-z}{e^{-\hat{N}}-z}\cdot\frac{e^{-\hat{x}}-z^{-1}}{e^{-\hat{N}}-z^{-1}}\right)^c.$$

Passing to the limit $\varepsilon\to0$ in the integrand of the contour integral (23.4), we get the result.   □

The next step is to show that the right-hand side of (23.2) as a function of c is the Laplace transform of a function; that is, there exists $h^\infty(\hat{x}, \hat{y})$ such that

$$\int_{-\infty}^{\infty} h^\infty(\hat{x}, \hat{y}) e^{-c\hat{y}} \, d\hat{y} = \frac{1}{2\pi i} \oint \mathcal{G}_{\hat{x}}(z)^c \frac{dz}{c^2 z}.$$

This function $h^\infty(\hat{x}, \hat{y})$ can be described quite exactly. Let us outline the computation of its derivative; for a more detailed argument, see the Proof of Corollary 6.3 in Ahn (2018). We begin by integrating by parts:

$$\int_{-\infty}^{\infty} \partial_{\hat{y}} h^\infty(\hat{x}, \hat{y}) e^{-c\hat{y}} \, d\hat{y} = c \int_{-\infty}^{\infty} h^\infty(\hat{x}, \hat{y}) e^{-c\hat{y}} \, d\hat{y} = \frac{1}{2\pi i} \oint \mathcal{G}_{\hat{x}}(z)^c \frac{dz}{cz}.$$
(23.6)

Define

$$\mu_{\hat{x}}(u) := \partial_{\hat{y}} h^\infty(\hat{x}, \hat{y})|_{e^{-\hat{y}}=u}.$$

The left-hand side of (23.6) can be rewritten as

$$\int_{-\infty}^{\infty} \partial_{\hat{y}} h^\infty(x, y) e^{-c\hat{y}} \, d\hat{y} = \int_{0}^{\infty} \mu_{\hat{x}}(u) u^c \frac{du}{u}.$$

Information on the moments of the measure $\mu_{\hat{x}}(u) du$ allows us to write down its Stieltjes transform:

$$\begin{aligned} S_{\mu_{\hat{x}}}(w) &:= \int_{0}^{\infty} \frac{\mu_{\hat{x}}(u) du}{w - u} = \sum_{c=1}^{\infty} w^{-c} \int_{0}^{\infty} \mu_{\hat{x}}(u) u^c \frac{du}{u} \\ &= \sum_{c=1}^{\infty} w^{-c} \frac{1}{2\pi i} \oint \mathcal{G}_{\hat{x}}(z)^c \frac{dz}{cz} = \frac{1}{2\pi i} \oint \ln\left(1 - \frac{\mathcal{G}_{\hat{x}}(z)}{w}\right) \frac{dz}{z}. \end{aligned}$$ (23.7)

Here, the fourth equality is obtained by interchanging the sum and integral and recognizing the internal expression as the power-series expansion of the logarithm. The parameter w should be chosen to be very large to guarantee the convergence of the sums, and the contour in the integral is as in Lemma 23.4 (i.e., it encloses 0 and $e^{-\hat{N}}$ but not $e^{\hat{N}}$).

The integrand has two singularities inside the integration contour: a simple pole at $z = 0$ and a more complicated singularity at the root of the equation $\mathcal{G}_{\hat{x}}(z) = w$ close to $e^{-\hat{N}}$ (because $\mathcal{G}_{\hat{x}}(z)$ has a simple pole at $e^{-\hat{N}}$, such a root exists and is unique for large w). Let us denote this root through z_0, and let γ be a contour enclosing z_0. Integrating by parts, we have

$$S_{\mu_{\hat{x}}}(w) = \ln\left(1 - \frac{\mathcal{G}_{\hat{x}}(0)}{w}\right) + \frac{1}{2\pi i}\oint_\gamma \ln\left(1 - \frac{\mathcal{G}_{\hat{x}}(z)}{w}\right)\frac{dz}{z}$$

$$= \ln\left(1 - \frac{\mathcal{G}_{\hat{x}}(0)}{w}\right) + \frac{1}{2\pi i}\oint_\gamma \frac{\mathcal{G}_{\hat{x}}'(z)}{\mathcal{G}_{\hat{x}}(z) - w}\ln(z)dz. \tag{23.8}$$

The last integral has a simple pole at z_0 with residue $\ln(z_0)$. Using also the explicit value for $\mathcal{G}_{\hat{x}}(0)$, we conclude that

$$S_{\mu_{\hat{x}}}(w) = \ln\left(1 - \frac{e^{\hat{N}}}{w}\right) + \ln(z_0), \tag{23.9}$$

where z_0 is still the root of the equation

$$w = \mathcal{G}_{\hat{x}}(z) \tag{23.10}$$

in a neighborhood of $e^{-\hat{N}}$.

By its definition, the Stieltjes transform $S_{\mu_{\hat{x}}}(w)$ is a complex analytic function of w outside of the (real) support of $\mu_{\hat{x}}$. Hence, we can analytically continue the formula (23.9) from large w to arbitrary w; the formula will remain the same, with the only difference being in the choice of the root of (23.10) – we no longer can guarantee that the desired root is the one closest to $e^{-\hat{N}}$.

After we figure out the Stieltjes transform, we can reconstruct the measure $\mu_{\hat{x}}(u)du$ itself. We note that when w crosses the real axis at a real point u, the imaginary part of $S_{\hat{x}}$ makes a jump[1] by $2\pi\mu_{\hat{x}}(u)$. We conclude that $2\pi\partial_{\hat{y}}h^\infty(\hat{x},\hat{y})$ is the jump of the imaginary part of (23.9) at the point $w = e^{-\hat{y}}$. Note that we expect the limit shape $h^\infty(\hat{x},\hat{y})$ to be identical to 0 for $\hat{y} \le -\hat{N}$ because of our choice of definition of the height function (alternatively, this can also be shown by analyzing the function $S_{\mu_{\hat{x}}}(w)$). Thus, we can further consider only $\hat{y} > -\hat{N}$. Then, for the first term in (23.9), we notice that $\frac{e^{\hat{N}}}{w} = e^{\hat{N}+\hat{y}} > 1$. Hence, the imaginary part of the logarithm makes a jump by 2π.

For the second term in (23.9), note that with the substitution of $w = e^{-\hat{y}}$ and using the definition of $\mathcal{G}_{\hat{x}}(z)$ from (23.3), we transform (23.10) into a quadratic equation in z with real coefficients. It either has real roots (in which case, there is no interesting jump for $\ln(z_0)$) or complex conjugate roots. One can check that the jump of $\ln(z_0)$ comes precisely from the choice of the root in the upper half-plane changing to the choice of the one in the lower half-plane. Hence, the jump of the imaginary part of $\ln(z_0)$ is twice the argument of z_0. Let us record our findings in a theorem.

---

[1] For simplicity, assume that u is a point of continuity of $\mu_{\hat{x}}$. Then this fact is proven by direct computation of the imaginary part as we approach u from the upper and lower half-planes.

Theorem 23.5 *Let $\zeta(\tilde{x}, \tilde{y})$ be a root of the quadratic equation*

$$\frac{1-z}{e^{-\hat{N}} - z} \cdot \frac{e^{-\hat{x}} - z^{-1}}{e^{-\hat{N}} - z^{-1}} = e^{-\hat{y}}, \tag{23.11}$$

chosen so that $\zeta(\hat{x}, \hat{y})$ is continuous in $(\hat{x}, \hat{y})$ and ζ is in the upper half-plane $\mathbb{H}$, if (23.11) has two complex conjugate roots. Then

$$\partial_{\hat{y}} h^{\infty}(\hat{x}, \hat{y}) = 1 - \frac{\arg \zeta(\hat{x}, \hat{y})}{\pi}.$$

Remark 23.6   Notice that if $\zeta(\tilde{x}, \tilde{y})$ is real, then $\partial_{\hat{y}} h^{\infty}(\hat{x}, \hat{y}) = 0$ or 1. This corresponds to the frozen region, either beneath the plane partition (where h is not changing) or above the plane partition (where the surface is vertical, and h increases with $\hat{y}$). For $(\hat{x}, \hat{y})$ such that $\zeta(\hat{x}, \hat{y}) \in \mathbb{H}$, we have

$$0 < \partial_{\hat{y}} h^{\infty}(\hat{x}, \hat{y}) < 1.$$

This corresponds exactly to the liquid region D. Therefore, the liquid region may be alternatively defined as the set of $(\tilde{x}, \tilde{y}) \in \mathbb{R}^2$ such that (23.11) has a pair of complex conjugate roots.

Remark 23.7   Let us match the result of Theorem 23.5 with the Kenyon–Okounkov theory developed in Lectures 9 and 10. We define a complex number z so that

$$\zeta(\hat{x}, \hat{y}) = e^{\hat{x}}(1 - \bar{z}(\hat{x}, \hat{y})).$$

Note that because ζ is in the upper half-plane, so is z. Conjugating the equation in (23.11) and rewriting it in terms of z, we get

$$Q(ze^y, (1-z)e^x) = 0, \qquad Q(u, v) = u(1 - v) - (e^{-\hat{N}} - v)(1 - ve^{-\hat{N}}).$$

This is precisely the equation of the form appearing in Theorems 9.8 and 10.1 and giving solutions to the complex Burgers equation for the limit shape. Indeed, the parameter c appearing in those theorems is -1 in our setting; see Corollary 9.6 and identify ε with $1/L$. Note also that $\arg(\zeta)$ is computing the density of horizontal $\diamond$. Hence, this density is also computed by $-\arg(1 - z)$, which matches Figure 9.1. With a bit more care, one can show that $z(\hat{x}, \hat{y})$ is precisely Kenyon–Okounkov's complex slope.

Exercise 23.8   Extend Theorems 23.1 and 23.5 to q^{Volume}-weighted plane partitions in an $N \times \lfloor \alpha N \rfloor$ rectangle, $\alpha > 0$, as $N \to \infty$.

Exercise 23.9   Sending $\hat{N} \to \infty$ in the result of Theorem 23.5, we get the limit shape for the plane partitions without a bounding rectangle. Find explicitly the formula for the boundary of the frozen region in this situation. The answer (drawn on the plane of panel (b) in Figure 22.1) should be invariant under rotations by 120 degrees.

The limit-shape theorem for q^{Volume}-weighted plane partitions (without bounding rectangle) was addressed by several authors, some of whom did not know about the existence of others. We refer the reader to Blöte and Hilhorst (1982), Nienhuis et al. (1984), and Vershik (1997) for the first appearances and to Cerf and Kenyon (2001) and Okounkov and Reshetikhin (2003) for later treatments in the mathematical literature.

23.2 Central Limit Theorem

Our next aim is to show that the centered height function

$$\sqrt{\pi} \left[h\left(\frac{\hat{x}}{\varepsilon}, \frac{\hat{y}}{\varepsilon} \right) - \mathbb{E}\, h\left(\frac{\hat{x}}{\varepsilon}, \frac{\hat{y}}{\varepsilon} \right) \right]$$

converges to the pullback of the GFF under the map $\zeta(\hat{x}, \hat{y})$ from Theorem 23.5.

We briefly revisit the definition of the GFF; see Lecture 11 for more details.

Definition 23.10   The GFF on the upper half-plane $\mathbb{H} = \{z \in \mathbb{C} \mid \Im(z) > 0\}$ is a random generalized centered Gaussian function GFF on $\mathbb{H}$ with covariance

$$\mathbb{E}\, GFF(z)GFF(w) = -\frac{1}{2\pi} \ln \left| \frac{z - w}{z - \bar{w}} \right|.$$

Given a bijection $\zeta : D \to \mathbb{H}$, where $D \subset \mathbb{R}^2$, we define the GFF pullback.

Definition 23.11   The ζ-pullback of the GFF is a random distribution $GFF \circ \zeta$ such that

$$\mathbb{E}[GFF \circ \zeta(x_1, y_1) \cdot GFF \circ \zeta(x_2, y_2)] = -\frac{1}{2\pi} \ln \left| \frac{\zeta(x_1, y_1) - \zeta(x_2, y_2)}{\zeta(x_1, y_1) - \bar{\zeta}(x_2, y_2)} \right|.$$

Theorem 23.12   *For the height functions of q^{Volume}-weighted plane partitions, we have*

$$\left\{ \sqrt{\pi} \sum_{y \in \varepsilon \mathbb{Z}} \varepsilon \left(h\left(\frac{\hat{x}}{\varepsilon}, \frac{\hat{y}}{\varepsilon} \right) - \mathbb{E}\, h\left(\frac{\hat{x}}{\varepsilon}, \frac{\hat{y}}{\varepsilon} \right) \right) e^{-c\hat{y}} \right\}_{c > 0, -\hat{N} \le \hat{x} \le \hat{N}}$$

$$\to \left\{ \iint_{D_{\hat{x}}} H \circ \zeta(\hat{x}, \hat{y}) e^{-cy} \right\}_{c > 0, -N \le x \le N} \qquad (23.12)$$

in the sense of the convergence of finite-dimensional distributions as $\varepsilon \to 0$.

Here, $\zeta(\hat{x}, \hat{y})$ is as in Theorem 23.5, D is the liquid region (which can be defined as those $(\hat{x}, \hat{y})$ for which ζ is nonreal), and $D_{\hat{x}}$ is the section of D by the vertical line with abscissa $\hat{x}$.

Remark 23.13 By Remark 23.7, $\overline{\zeta} = e^x(1 - z)$, where z is the Kenyon–Okounkov complex slope. Because $Q(e^y z, e^x(1 - z)) = 0$ for the analytic Q, the complex structure of $\overline{\zeta}$ is the same as the one of $e^y z$. Noting also that the Laplacian in the definition of the GFF is unchanged under the complex conjugation of the complex structure, we conclude that Theorem 23.12 proves a particular case of Conjecture 11.1.

Exercise 23.14 Show that $\zeta \colon D \to \mathbb{H}$ is a homeomorphism from D to $\mathbb{H}$, which maps sets of the form $\{(x, y) \in D : y \in \mathbb{R}\}$ for fixed x to half circles in $\mathbb{H}$ with real center points.

The convergence in the sense of finite-dimensional distributions in Theorem 23.12 means that if we take any finite collection

$$c_1, \dots, c_k, \quad -\hat{N} \le \hat{x}_1, \dots, \hat{x}_k \le \hat{N},$$

then we have the convergence in the distribution for random vectors

$$\lim_{\varepsilon \to 0} \left(\sum_{\hat{y} \in \varepsilon \mathbb{Z}} \varepsilon \left(h\left(\frac{\hat{x}_i}{\varepsilon}, \frac{\hat{y}}{\varepsilon}\right) - \mathbb{E}\, h\left(\frac{\hat{x}_i}{\varepsilon}, \frac{\hat{y}}{\varepsilon}\right)\right) e^{-c_i \hat{y}} \right)^k_{i=1}$$

$$= \left(\int_{D_{\hat{x}_i}} H \circ \zeta(\hat{x}, \hat{y}) e^{-c_i \hat{y}} d\hat{y} \right)^k_{i=1}. \quad (23.13)$$

We can prove such a statement by showing that these random vectors converge to a Gaussian limit and then matching their covariance. We outline some of the ideas of the proof here and refer the reader to Ahn (2018) for more details.

Definition 23.15 Given random variables $X_1, \dots, X_n$, their *joint mixed cumulant* $\kappa(X_1, \dots, X_n)$ is defined as

$$\kappa(X_1, \dots, X_n) = \frac{\partial^n}{\partial t_1 \cdots \partial t_n} \ln \mathbb{E}\left[\exp\left(\mathbf{i} \sum_{i=1}^{N} t_i X_i \right) \right]\Bigg|_{t_1 = t_2 = \cdots = t_n = 0}.$$

If we fix random variables $Y_1, \dots, Y_k$, then its family of cumulants is defined as all possible $\kappa(X_1, \dots, X_n)$, $n = 1, 2, \dots$, where X_i coincide with various Y_j with possible repetitions. n is then the order of the cumulant. For instance, if we have one random variable $Y_1 = Y$, then its first cumulant is $\kappa(Y) = \mathbb{E}\, Y$, and the second cumulant $\kappa(Y, Y) = \mathbb{E}\, Y^2 - (\mathbb{E}\, Y)^2$. For two random variables, the second-order cumulant is the covariance, $\kappa(X_1, X_2) = \text{cov}(X_1, X_2)$.

For a Gaussian vector $(Y_1, \ldots, Y_k)$, the logarithm of its characteristic function is a second-degree polynomial. This leads to the following statement:

Exercise 23.16   A random vector $(Y_1, \ldots, Y_k)$ has a Gaussian distribution if and only if all joint cumulants of its coordinates of order ≥ 3 vanish.

In principle, one can express joint cumulants through joint moments. In this way, vanishing of the higher-order cumulants is equivalent to Wick's formula for the computation of joint moments for a Gaussian vector as a combinatorial sum involving perfect matchings.

Hence, in order to prove Theorem 23.12, we need two steps:

1 Compute the covariance for random variables $\sum_{\hat{y} \in \varepsilon \mathbb{Z}} \varepsilon \left(h \left(\frac{\hat{x}_i}{\varepsilon}, \frac{\hat{y}}{\varepsilon} \right) \right) e^{-c_i \hat{y}}$ using Theorem 22.7 and show that it converges to the covariance of the vectors in the right-hand side of (23.13).

2 Compute higher-order cumulants for the same random variables and show that they converge to 0.

We only outline the first step. The asymptotic vanishing of the higher-order cumulants for the second step holds in much greater generality for random variables, whose joint moments are given by contour integrals, as in Theorem 22.7. We refer the reader to Section 4.3 in Borodin and Gorin (2015), Section 6.2 in Gorin and Zhang (2018), and Section 5.3 in Ahn (2018) for such arguments. For the identification of the covariance structure, take $c_1, c_2 > 0$ and $-\hat{N} \leq \hat{x}_1, \hat{x}_2 \leq \hat{N}$. Then

$$
\mathrm{Cov}\left(\sum_{\hat{y} \in \varepsilon \mathbb{Z}} \varepsilon \left(h\left(\frac{\hat{x}_1}{\varepsilon}, \frac{\hat{y}}{\varepsilon} \right) - \mathbb{E}\, h\left(\frac{\hat{x}_1}{\varepsilon}, \frac{\hat{y}}{\varepsilon} \right) \right) e^{-c_1 \hat{y}}, \right.
$$

$$
\left. \sum_{\hat{y} \in \varepsilon \mathbb{Z}} \varepsilon \left(h\left(\frac{\hat{x}_2}{\varepsilon}, \frac{\hat{y}}{\varepsilon} \right) - \mathbb{E}\, h\left(\frac{\hat{x}_2}{\varepsilon}, \frac{\hat{y}}{\varepsilon} \right) \right) e^{-c_2 \hat{y}} \right)
$$

$$
= \mathbb{E}\left[\prod_{i=1}^{2} \sum_{\hat{y} \in \varepsilon \mathbb{Z}} \varepsilon h\left(\frac{\hat{x}_i}{\varepsilon}, \frac{\hat{y}}{\varepsilon} \right) e^{-c_i \hat{y}} \right] - \prod_{i=1}^{2} \mathbb{E}\left[\varepsilon \sum_{\hat{y} \in \varepsilon \mathbb{Z}} h\left(\frac{\hat{x}_i}{\varepsilon}, \frac{\hat{y}}{\varepsilon} \right) e^{-c_i \hat{y}} \right].
$$

By Theorem 22.7, the previous equation is

$$
\frac{1}{(2\pi \mathbf{i})^2} \oint \oint \left(\frac{(z_2 - \frac{t_1}{t_2} z_1)(z_2 - z_1)}{(z_2 - t_1 z_1)(z_2 - \frac{1}{t_2} z_1)} - 1 \right) \prod_{i=1}^{2} \left[G_{x_i}(z_i; t_i) \frac{\varepsilon \, dz_i}{(t_i^{-1} - 1)(1 - t_i)z_i} \right],
$$

where $t_i = e^{-c_i \varepsilon}$, $x_i = \frac{\tilde{x}_i}{\varepsilon}$. Because

$$\frac{(z_2 - \frac{t_1}{t_2} z_1)(z_2 - z_1)}{(z_2 - t_1 z_1)(z_2 - \frac{1}{t_2} z_1)} - 1 = \frac{(1 - t_1)(\frac{1}{t_2} - 1) z_1 z_2}{(z_2 - t_1 z_1)(z_2 - \frac{1}{t_2} z_1)},$$

the ε factors balance out with $1 - t_i$ to give us an overall $1/(c_1 c_2)$ factor. By (23.5), we obtain

$$\frac{1}{c_1 c_2} \cdot \frac{1}{(2\pi i)^2} \oint \oint \frac{\mathcal{G}_{\hat{x}_1}(z_1)^{c_1} \mathcal{G}_{\hat{x}_2}(z_2)^{c_2}}{(z_2 - z_1)^2} \, dz_1 \, dz_2. \tag{23.14}$$

It remains to show that (23.14) is the covariance of the GFF paired with exponential test functions on the sections.

Let C_i be a closed contour symmetric with respect to the x-axis, and the part of C_i in $\mathbb{H}$ is the half-circle that is the ζ-image of the vertical section of the liquid region with abscissa $\hat{x}_i$ (i.e., the set $\{\zeta(\hat{x}_i, \hat{y}) \mid y \in D_{\hat{x}_i}\}$). We deform the integration contours in (23.5) to C_i and then split $C_i = C_i^+ \cup C_i^-$, where $C_i^+ = C_i \cap \mathbb{H}$. Changing variables via $\zeta(x, y) \mapsto (x, y)$ and using (23.11), we rewrite (23.14) as

$$-\frac{1}{4\pi^2 c_1 c_2} \int_{D_{\hat{x}_2}} \int_{D_{\hat{x}_1}} \frac{e^{-c_1 \hat{y}_1} e^{-c_2 \hat{y}_2}}{(\zeta(\hat{x}_1, \hat{y}_1) - \zeta(\hat{x}_2, \hat{y}_2))^2} \frac{\partial \zeta}{\partial \hat{y}_1}(\hat{x}_1, \hat{y}_1) \frac{\partial \zeta}{\partial \hat{y}_2}(\hat{x}_2, \hat{y}_2) \, d\hat{y}_1 \, d\hat{y}_2$$

$$+\frac{1}{4\pi^2 c_1 c_2} \int_{D_{\hat{x}_2}} \int_{D_{\hat{x}_1}} \frac{e^{-c_1 \hat{y}_1} e^{-c_2 \hat{y}_2}}{(\zeta(\hat{x}_1, \hat{y}_1) - \overline{\zeta}(\hat{x}_2, \hat{y}_2))^2} \frac{\partial \zeta}{\partial \hat{y}_1}(\hat{x}_1, \hat{y}_1) \frac{\partial \overline{\zeta}}{\partial \hat{y}_2}(\hat{x}_2, \hat{y}_2) \, d\hat{y}_1 \, d\hat{y}_2$$

$$+\frac{1}{4\pi^2 c_1 c_2} \int_{D_{\hat{x}_2}} \int_{D_{\hat{x}_1}} \frac{e^{-c_1 \hat{y}_1} e^{-c_2 \hat{y}_2}}{(\overline{\zeta}(\hat{x}_1, \hat{y}_1) - \zeta(\hat{x}_2, \hat{y}_2))^2} \frac{\partial \overline{\zeta}}{\partial \hat{y}_1}(\hat{x}_1, \hat{y}_1) \frac{\partial \zeta}{\partial \hat{y}_2}(\hat{x}_2, \hat{y}_2) \, d\hat{y}_1 \, d\hat{y}_2$$

$$-\frac{1}{4\pi^2 c_1 c_2} \int_{D_{\hat{x}_2}} \int_{D_{\hat{x}_1}} \frac{e^{-c_1 \hat{y}_1} e^{-c_2 \hat{y}_2}}{(\overline{\zeta}(\hat{x}_1, \hat{y}_1) - \overline{\zeta}(\hat{x}_2, \hat{y}_2))^2} \frac{\partial \overline{\zeta}}{\partial \hat{y}_1}(\hat{x}_1, \hat{y}_1) \frac{\partial \overline{\zeta}}{\partial \hat{y}_2}(\hat{x}_2, \hat{y}_2) \, d\hat{y}_1 \, d\hat{y}_2. \tag{23.15}$$

Integrate by parts on $\hat{y}_1$ and $\hat{y}_2$ for each summand, observing that the boundary terms cancel because the value of $\zeta(x, \cdot)$ at the end points of D_x is real, to obtain

$$\frac{-1}{4\pi^2} \int_{D_{\hat{x}_2}} \int_{D_{\hat{x}_1}} e^{-c_1 \hat{y}_1 - c_2 \hat{y}_1} \Big[\ln(\zeta(x_1, y_1) - \zeta(x_2, y_1)) - \ln(\zeta(x_1, y_1) - \overline{\zeta}(x_2, y_1))$$

$$- \ln(\overline{\zeta}(x_1, y_1) - \zeta(x_2, y_1)) + \ln(\overline{\zeta}(x_1, y_1) - \overline{\zeta}(x_2, y_1)) \Big] \, dy_1 \, dy_1$$

$$= -\frac{1}{2\pi^2} \int_{D_{\hat{x}_2}} \int_{D_{\hat{x}_1}} e^{-c_1 \hat{y}_1} e^{-c_2 \hat{y}_1} \ln \left| \frac{\zeta(\hat{x}_1, \hat{y}_1) - \zeta(\hat{x}_2, \hat{y}_1)}{\zeta(\hat{x}_1, \hat{y}_1) - \overline{\zeta}(\hat{x}_2, \hat{y}_1)} \right| \, d\hat{y}_1 \, d\hat{y}_2.$$

This is precisely the desired formula for the covariance of the integrals of the GFF against exponential test functions. The prefactor $\frac{1}{2\pi^2}$ matches the one in Conjecture 11.1; it also matches the ratio of $\frac{1}{2\pi}$ in Definition 23.11 and $(\sqrt{\pi})^2$ coming from the prefactor in (23.12).

24

Lecture 24: Discrete Gaussian Component in Fluctuations

24.1 Random Heights of Holes

There are two frameworks for picking a uniformly random tiling:

1. Fix a domain, and pick a tiling of this domain uniformly at random.

2. Fix a domain and a height function on the boundary, and pick a height function inside the domain extending the boundary height function uniformly at random.

For simply connected domains, the two points of view are equivalent. For domains with holes, the two might not be equivalent because the heights of holes are fixed in the second framework, but there can be multiple possibilities for the heights of the holes in the first framework. So in the first framework the height of a hole is generally a nonconstant random variable. In expectation, the height of a hole grows linearly with the size of the domain. In this lecture, we are interested in determining the order of fluctuations of the height around its expectation. It turns out that the perspective that leads to the correct answer is thinking of these as global fluctuations and expecting the fluctuations to have finite order, as the Gaussian free field (GFF) heuristic predicts (cf. Lectures 11, 12, 21, and 23). The limit is a discrete random variable, and we will present two approaches to finding it. The first is entirely heuristic but applies to very general domains. The second approach can be carried out rigorously (although our exposition here will still involve some heuristic steps) but only applies to a specific class of domains.

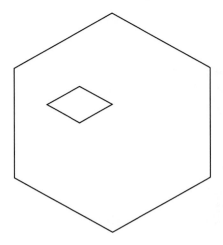

Figure 24.1 A hexagon with a rhombic hole.

24.2 Discrete Fluctuations of Heights through GFF Heuristics

In this section, we are going to work with an arbitrary domain with a single hole (which can be of any shape), and a reader can take a hexagon with a rhombic hole in Figures 24.1, and 24.2 as a running example. Let h be the *centered* height function on the domain. By this, we mean that at each point, h is the difference between the height function and the expected height at that point. Throughout this section, we work with a modified version of the height function of (9.3) and (9.4). This means that $h(x, y)$ can be thought of as the (centered) total number of horizontal lozenges $\diamond$ in the tiling situated on the same vertical line as (x, y) below (x, y).

We start with the GFF heuristic; see Lectures 11 and 12 for the derivation. For large domains, the distribution of the centered height function (or rather, its density) can be approximated as

$$\mathbb{P}(h) \approx \exp\left(-\frac{\pi}{2}\iint \|\nabla h\|^2\right).$$

This integral is taken in the complex coordinates related to the GFF asymptotics; the integration domain is in bijection with the liquid region.

Outside the liquid region, the lozenges are deterministic with overwhelming probability. This means that for a typical tiling, the height increment between any two points in a frozen region is deterministic and equal to the increment in expectation. Hence, the centered height function is the same at all points of

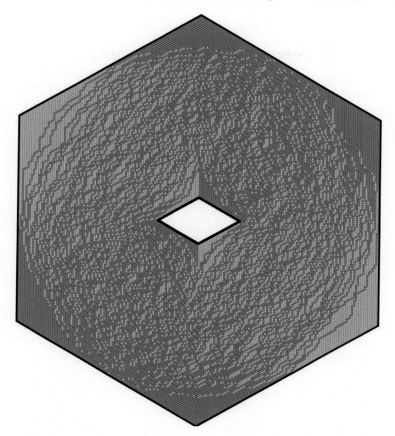

Figure 24.2 A typical tiling of a hexagon with a hole. (Simulation by Leonid Petrov.)

a frozen region. Thus, picking a point on the outer boundary of the domain to have zero centered height, we have that the centered height function on the outer frozen region is zero, and the centered height on the frozen region around the hole is δ, which is the same as the fluctuation $h(B)$[1] of the height of the hole. Using our GFF heuristic,

$$\mathbb{P}(h(B) = \delta) \approx \int_{\Omega_\delta} \exp\left(-\frac{\pi}{2} \iint \|\nabla h\|^2\right), \qquad (24.1)$$

---

[1] $h(B)$ stands for "height of boundary."

where Ω_δ is the subset of height functions for which the height of the inner frozen region is δ.

We would like to understand how (24.1) changes as we vary δ. For that, we take a (unique) harmonic function g_δ on the liquid region that satisfies the $g_\delta = 0$ boundary condition on the outer boundary and the $g_\delta = \delta$ boundary condition on the internal boundary surrounding the hole. The harmonicity is with respect to the local coordinates of the GFF complex structure. Then for $h \in \Omega_\delta$, we can define $\tilde{h} = h - g_\delta$, and notice that $\tilde{h}$ is 0 on *all* boundaries of the liquid region. This takes us a step closer to evaluating the integral:

$$h = \tilde{h} + g_\delta,$$

$$\iint \|\nabla h\|^2 = \iint \|\nabla \tilde{h} + \nabla g_\delta\|^2 = \iint \|\nabla \tilde{h}\|^2 + \iint \|\nabla g_\delta\|^2 + 2 \iint \langle \nabla \tilde{h}, \nabla g_\delta \rangle.$$

We can integrate the last summand by parts:

$$\iint \langle \nabla \tilde{h}, \nabla g_\delta \rangle = \text{boundary term} - \iint \tilde{h} \nabla^2 g_\delta.$$

The boundary term is 0 because $\tilde{h}$ is 0 on the boundary, and the other term is 0 because $\nabla^2 g_\delta = 0$ as g_δ is harmonic.

Hence, we can write the probability we are interested in as

$$\mathbb{P}(h(B) = \delta) \approx \int_{\Omega_\delta} \exp\left(-\frac{\pi}{2} \iint \|\nabla \tilde{h}\|^2 + \iint \|\nabla g_\delta\|^2\right).$$

Notice that for any δ, δg_1 is harmonic and satisfies the same boundary conditions as g_δ, so $g_\delta = \delta g_1$. Hence, we can write

$$\mathbb{P}(h(B) = \delta) \approx \int_{\Omega_\delta} \exp\left(-\frac{\pi}{2} \iint \|\nabla \tilde{h}\|^2\right) \cdot \exp\left(-\frac{\pi}{2} \delta^2 \iint \|\nabla g_1\|^2\right)$$

$$= \exp(-C\delta^2) \int_{\Omega_\delta} \exp\left(-\frac{\pi}{2} \iint \|\nabla \tilde{h}\|^2\right).$$

Note that $\tilde{h}$ in the last integral is an element of Ω_0 and therefore is independent of δ. Hence, the integral evaluates to a δ-independent constant. Therefore, we have provided a heuristic argument for the following conjecture:

Conjecture 24.1   *As the linear size of the domain $L \to \infty$, the height H of a hole is an integer random variable that becomes arbitrarily close in distribution to the* discrete Gaussian distribution

$$const \cdot \exp(-C(H - m)^2), \tag{24.2}$$

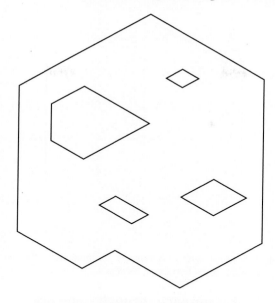

Figure 24.3 A domain with multiple holes.

where m is a certain (unknown at this point) shift, and the scaling factor C can be computed as the Dirichlet energy

$$C = \frac{\pi}{2} \iint \|\nabla g_1\|^2,$$

and g_1 is the harmonic function in the liquid region (with respect to the Kenyon–Okounkov complex structure) with $g_1 = 0$ on the external boundary and $g_1 = 1$ on the internal boundary.

The same heuristic argument can also be given for domains with multiple holes, as in Figure 24.3, leading us to the following conjecture, which the author learned from Slava Rychkov:

Conjecture 24.2 *For a domain with an arbitrary collection of K holes, the asymptotic distribution of the hole heights is a K-dimensional discrete Gaussian vector with a scale matrix given by the quadratic form $\iint \langle \nabla f, \nabla g \rangle$ of harmonic functions in the liquid region with a Kenyon–Okounkov complex structure and with prescribed boundary conditions.*

We are not aware of any simple formula for the shifts m, and this is an interesting open question.

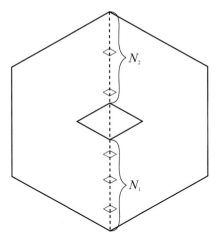

Figure 24.4 A hexagon with a symmetric rhombic hole, the vertical line though its center, and some horizontal lozenges.

24.3 Approach through Log-Gases

In this section we consider the case of a regular hexagon with side length $N + D$ with a rhombic hole with side length D in the middle. We will take some heuristic steps, although the arguments can be carried out rigorously with more effort.

The height of the hole is determined by the lozenges on the vertical line that goes through the middle of the hexagon. Specifically, the height is determined by the number of horizontal lozenges that lie on the vertical line below the hole. Let this number be N_1, and let the number of horizontal lozenges on this line above the hole be N_2. Note that the geometry of the hexagon (the height function on the boundary) dictates that the total number of horizontal lozenges on this vertical line is N. So $N_1 + N_2 = N$. The fluctuation of the height of the hole is the same as the fluctuation of N_1.

Because we are looking for the distribution of N_1, it would be great to know the probability distribution for the positions of these horizontal lozenges. Fortunately, this was found in Lecture 19 for the hexagon, and essentially the same argument (which is omitted) also works for the hexagon with a hole. With $x_1 < x_2 < \cdots < x_N$ being the vertical coordinates of the N horizontal lozenges,

$$\mathbb{P}(x_1, x_2, \ldots, x_N) \propto \prod_{i<j} (x_i - x_j)^2 \prod_{i=1}^{N} w(x_i). \tag{24.3}$$

In our case, w is the product of several Pochhammer symbols; see Section 9.2 in Borodin et al. (2017).

Exercise 24.3 Extending the argument that produced (19.1), compute w explicitly.

Now, let us define a "potential" V by

$$w(x) = \exp\left(-NV\left(\frac{x}{N}\right)\right).$$

Given a position vector $\boldsymbol{x} = (x_1, x_2, \ldots, x_N)$, we also define the following probability measure:

$$\mu_N = \frac{1}{N}\sum_{i=1}^{N}\delta_{x_i/N}. \tag{24.4}$$

Because $\boldsymbol{x}$ is a random variable, μ_N is a random distribution. In the limit $N \to \infty$, μ_N becomes a continuous probability measure (note that because adjacent x_is are separated at least by distance 1, the density of the limiting measure is upper-bounded by 1). With this notation, we can rewrite

$$\mathbb{P}(\boldsymbol{x}) \propto \exp\left(N^2 I[\mu_N]\right),$$

where

$$I[\mu_N] = \iint_{x \neq y} \ln|x - y|\mu_N(dx)\mu_N(dy) - \int V(x)\mu_N(dx).$$

As $N \to \infty$, the measure μ_N concentrates near the maximizer of $I[\cdot]$; this is another face of the limit-shape theorem for random tilings; see Borodin et al. (2017).

Letting μ be the probability measure of density of at most 1 that maximizes $I[\cdot]$, we can consider a deviation $\frac{1}{N}g_N$ from the maximum (here, $\frac{1}{N}g_N$ is the difference of two probability measures, i.e., has total mass zero) and expand as follows:

$$I[\mu_N] = I\left[\mu + \frac{1}{N}g_N\right] = I[\mu]$$
$$+ \frac{1}{N}\left(2\iint \ln|x - y|\mu(dy)g_N(dx) - \int V(x)g_N(dx)\right)$$
$$+ \frac{1}{N^2}\ln|x - y|g_N(dx)g_N(dy).$$

Because μ is a maximizer and the middle term is a derivative of $I[\cdot]$ at μ (in the g_N direction), it vanishes. Thus, plugging back into the equation for $\mathbb{P}(\boldsymbol{x})$, we get the following proposition:

Proposition 24.4   *The law of fluctuations g_N as $N \to \infty$ becomes*

$$\mathbb{P}(g_N) \propto \exp\left(\iint \ln|x - y| g_N(dx) g_N(dy)\right).$$

This is an exponential of a quadratic integral, so we get a Gaussian law.

This proposition can be proved rigorously using the loop (Nekrasov) equations of Lecture 21; see Borodin et al. (2017), and see Borot and Guionnet (2013) and Johansson (1998) for a continuous-space version.

We now understand the distribution of the fluctuations g_N. Note that the support of g_N is the part of the vertical line in the liquid region, that is, a pair of intervals $[a, b] \cup [c, d]$, as in Figure 24.5. We would like to also understand the fluctuation of the height of the hole, which is equal to the fluctuation of $g_N[a, b] = -g_N[c, d]$. The distribution of the latter is

$$\mathbb{P}(g_N[a, b] = -g_N[c, d] = \delta) \propto \int_{\Omega_\delta} \exp\left(\iint \ln|x - y| g_N(dx) g_N(dy)\right). \tag{24.5}$$

Here, Ω_δ is the subset of all g_N with $g_N[a, b] = -g_N[c, d] = \delta$.

Lemma 24.5   *The expression in (24.5) is the same as*

$$\mathbb{P}(g_N[a, b] = -g_N[c, d] = \delta)$$

$$\propto \exp\left(\max_{g_N \text{ such that } g_N[a,b]=\delta} \iint \ln|x - y| g_N(dx) g_N(dy)\right)$$

$$= \exp\left(\iint \ln|x - y| g^\delta(dx) g^\delta(dy)\right),$$

where g^δ is defined to be the maximizer.

Proof   This is a general statement about Gaussian vectors, which we leave as an exercise: the density of projection of a high-dimensional vector onto a smaller subspace can be found through the maximization procedure.   □

In analogy with the argument from before, we again have $g^\delta = \delta g^1$, so we conclude with the following result:

Conjecture 24.6   *The height of a hole is asymptotically a discrete Gaussian (integer-valued) random variable; that is, it has the following distribution:*

$$const \cdot \exp(-C(H - m)^2),$$

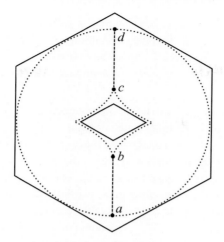

Figure 24.5 The dotted lines are the boundaries of frozen regions, and the dashed segments are the support of g_N.

where m is a certain shift, and the scale parameter C is given by the extremum of the logarithmic energy:

$$C = -\max \iint \ln|x - y| g_N(dx) g_N(dy),$$

with the maximum taken over all g_N on $[a, b] \cup [c, d]$, such that $g_N[a, b] = 1$, $g_N[c, d] = -1$.

Conjecture 24.6 can be proven rigorously, although with significant technical effort; see a forthcoming article of Borot et al.[2] The same argument can also be carried out for multiple symmetric holes on a vertical line.

24.4 Two-Dimensional Dirichlet Energy and One-Dimensional Logarithmic Energy

The material in this subsection is based on private communications of the author with Alexei Borodin, Slava Rychkov, and Sylvia Serfaty.

We have seen in the previous two sections two approaches that both gave discrete Gaussian distributions for the height function of the hole. However, the constants C in the exponents in Conjectures 24.1 and 24.6 have different expressions. For the former, C is given as the Dirichlet energy of a harmonic

---

[2] G. Borot, V. Gorin, and A. Guionnet, Fluctuations for multi-cut discrete β-ensembles and application to random tilings.

function with prescribed boundary conditions, which is equal to the minimal Dirichlet energy with these boundary conditions. For the latter, C is given by a log-energy minimization on the line. The two visually different expressions have to be equivalent, and in this section we provide a mathematical explanation for this.

We start by repeating and extending the statements of Conjectures 24.1 and 24.6 for the setting of $k \geq 1$ holes in the domain.

Recall that we would like to understand the discrete component of the global fluctuations in the uniformly random tilings of the domains with k holes. Figure 24.2 shows a sample of lozenge tilings in the one-hole situation, and Figure 24.6 shows a sample of *domino* tiling in the two-hole situation. Asymptotically, the discrete component should be given by discrete Gaussian random variables, which are random variables on $\mathbb{Z}^{k-1}$ depending on two sets of parameters: shift and scale matrix ("covariance," although the exact covariance of the coordinates is slightly different). In the case of one hole, this random variable is one-dimensional and has the weight

$$\frac{1}{Z} \exp\left(-\frac{(x-m)^2}{2s^2}\right), \quad x \in \mathbb{Z}. \tag{24.6}$$

We would like to compare two different ways of computing s^2 (and more generally, the scale matrix): the first one proceeds through the two-dimensional variational principle and its approximation by the Dirichlet energy in the appropriate coordinate system; the second one relies on the identification of a section of the tilings with 1D log-gas and the variational principle for the latter. The two ways generalize those in the previous subsections.

Let us start with the **first approach**.

We know from the variational principle that the number of tilings with a height approximating a height profile $h(x, y)$ can be approximately computed as

$$\exp\left(L^2 \int\int S(\nabla h) dx dy + o(L^2)\right), \tag{24.7}$$

where L is the linear size of the system, and the integration goes over the tiled domain (rescaled to be finite). As in Lecture 12, (24.7) can be used to give a quadratic approximation of the energy near the maximizer of (24.7), known as the limit shape. Denoting $\tilde{h} = h - \mathbb{E}h$, we get the law

$$\exp\left(-\frac{\pi}{2} \int\int |\nabla_\Omega(\tilde{h})|^2 dz\right). \tag{24.8}$$

The integral in (24.8) goes over the *liquid region* in (x, y)-coordinates because there are no fluctuations outside. The map Ω transforms the liquid region into a certain Riemann surface of the same topology; in the applications, this surface can be taken to be a domain in $\mathbb{C}$ through the Riemann uniformization, as in Theorem 11.13. The meaning of (24.8) is that we need to introduce a map or change of coordinates $(x, y) \mapsto \Omega(x, y)$ in order to turn the quadratic variation of (24.7) near its maximum into the Dirichlet energy. For our purposes, the map $\Omega(x, y)$ is defined only up to compositions with conformal functions because the latter leave the Dirichlet energy invariant.

The discrete Gaussian component arises as the integral of (24.8) over varying boundary conditions. A basic property of the Dirichlet energy in a domain is that for the sum of two functions $f = u+v$, such that v is harmonic in the domain and u vanishes on the boundary of the domain, the Dirichlet energy of f is the sum of Dirichlet energies of u and v. Hence, we can identify the integrals of (24.8) with different boundary conditions. Therefore, the scale of the Gaussian component can now be reconstructed through the computation of the Dirichlet energy of the harmonic functions (in the liquid region, in Ω-coordinate), with appropriate boundary conditions reflecting the height differences on different connected components of the boundary. For example, for the holey hexagon of Figure 24.4 there are two connected components of the boundary, and therefore, there are two heights. By recentering, we can assume that the height of the outer component is zero, and therefore, only one variable remains.

It is shown in Bufetov and Gorin (2019) that for a class of domains (gluing of trapezoids in the lozenge case and gluings of Aztec rectangles in the domino case), the map $\Omega(x, y)$ can be chosen so that it maps the liquid region into the complex plane with cuts. For instance, in the holey hexagon case of Figure 24.4, the cuts can be identified with intervals on the vertical symmetry axis on the picture; their end points coincide with the intersection of the frozen boundary of the liquid region with the symmetry axis; these are $[a, b]$ and $[c, d]$ in Figure 24.5. In the Aztec diamond with two holes shown in Figure 24.6, the cuts are on the horizontal axis going through the centers of the holes; again, the end points of the cuts lie on the frozen boundary.

We thus are led to the problem of finding the Dirichlet energy of harmonic functions in a domain with cuts. Let us now formulate the setting of the last problem in a self-contained form.

Take $2k$ real numbers $a_1 < b_1 < a_2 < \cdots < a_k < b_k$. We refer to intervals $[a_k, b_k]$ as *cuts*. Let $\mathbb{H}$ denote the (open) upper half-plane, and let $\bar{\mathbb{H}}$ denote the lower half-plane. Set $\mathbb{D} = \mathbb{H} \cup \bar{\mathbb{H}} \cup_{i=1}^{k} (a_i, b_i)$.

Fix k real numbers $n_1, \ldots, n_k$ subject to the condition $n_1 + \cdots + n_k = 0$. These numbers are fluctuations of the height function in the setting of the previous

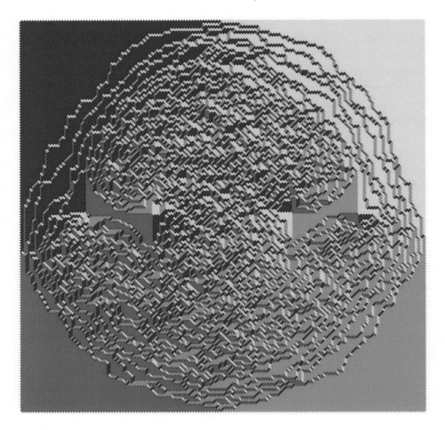

Figure 24.6 Domino tiling of Aztec diamond with two holes; there are two types of horizontal and vertical lozenges (due to checkerboard coloring) and, hence, four colors in the picture. (Drawing by Sevak Mkrtchyan.)

text: they are increments of the height as we cross a connected component of the liquid region (cf. N_1 and N_2 in Figure 24.4) with subtracted expectations. The zero-sum condition corresponds to the deterministic full increment of the height function between the points of the outer boundary. In particular, $k = 2$ in Figures 24.2 and 24.4, and $k = 3$ in Figure 24.6.

We consider *continuous* functions $h(z)$, $z = x + \mathbf{i}y$, in $\mathbb{C}$, differentiable in $\mathbb{D}$, and subject to the following boundary conditions on the real axis:

$$h(-\infty) = h(+\infty) = 0, \tag{24.9}$$

$$\frac{\partial}{\partial x} h(x) = 0 \text{ on } [-\infty, a_1] \cup [b_1, a_2] \cup \cdots \cup [b_{k-1}, a_k] \cup [b_k, +\infty], \tag{24.10}$$

$$h(b_i) - h(a_i) = n_i, \quad i = 1, \ldots, k. \tag{24.11}$$

We are interested in the Dirichlet energy of such functions,

$$\mathcal{E}(h) = \iint_\mathbb{D} |\nabla h|^2 dx dy = \iint_\mathbb{D} [(h_x)^2 + (h_y)^2] dx dy.$$

We further define a quadratic form C through

$$C : (n_1, \ldots, n_k) \to \frac{\pi}{2} \min_h \mathcal{E}(h), \text{ subject to } (24.9), (24.10), \text{and}(24.11).$$
$$\tag{24.12}$$

Note that by the Dirichlet principle, the minimizer in (24.12) is a harmonic function in $\mathbb{D}$.

Claim 24.7 *The quadratic form C is precisely the quadratic form for the discrete Gaussian component of the fluctuations of the height function in random tilings in Conjecture 24.1.*

From the **second perspective**, we saw in the previous section that the section of a random tiling along a singled-out axis gives rise to a discrete log-gas. This axis is the vertical symmetry axis in Figures 24.2 and 24.4 and the horizontal axis going through the centers of the holes in Figure 24.6. Other examples can be found in Borodin et al. (2017) and Bufetov and Gorin (2019). Clearly, it is enough to study what is happening on this axis in order to find the desired discrete Gaussian component of the fluctuation.

In more detail, in Figure 24.4, the distribution of the horizontal lozenges on the vertical section of the tiling by the symmetry axis of the domain has the form of the log-gas. The particles x_i in (24.3) belong to two segments – below and above the hole. We are thus interested in fluctuations of the number of particles in one of the segments. For a large class of tilings, we will have a generalization of (24.3), with particles now confined to k segments instead of 2.

If we denote through μ_N the empirical measure of the particles as in (24.4), then (24.3) can be rewritten (ignoring the diagonal) as follows:

$$\exp\left[N^2 \left(\iint \ln |x - y| \mu_N(dx) \mu_N(dy) + \int \ln w(Nx) \mu(dx) \right) \right]. \tag{24.13}$$

The maximizer of the functional in the exponent of (24.13) is the equilibrium measure. The fluctuations of the numbers of particles (called "filling fractions" in this context) can be thus obtained by varying this functional near its extremum. The rigorous justification of this procedure is the subject of a forthcoming article of Borot et al.[3] In the continuous setting (when x_i in (24.3) are real numbers rather than integers), a detailed rigorous analysis for an arbitrary number of cuts k was done in Borot and Guionnet (2013).

Let us describe how the approximation of the functional near the equilibrium measure looks. We assume that the *bands* of the equilibrium measure are k intervals $[a_1, b_1], \ldots, [a_k, b_k]$. Recall that bands are regions where the density is not zero and not one as in the liquid region for tilings (i.e., not frozen/saturated, the latter corresponding to the areas in Figures 24.2 and 24.4 that are densely packed with particles).

Given $n_1, \ldots, n_k$ with $n_1 + \cdots + n_k = 0$, we are led to consider *signed* measures ν on the union of the intervals $[a_i, b_i]$ subject to filling fractions:

$$\nu([a_i, b_i]) = n_i, \quad i = 1, \ldots, k. \tag{24.14}$$

We consider the logarithmic energy of such measure given by

$$\tilde{\mathcal{E}} = -\int\int \ln|x - y|\nu(dx)\nu(dy). \tag{24.15}$$

We define a quadratic functional $\tilde{C}$ through

$$\tilde{C}: (n_1, \ldots, n_k) \to \min_{\nu} \tilde{\mathcal{E}}(\nu) \text{ subject to (24.14).} \tag{24.16}$$

Claim 24.8   *The quadratic form $\tilde{C}$ is precisely the quadratic form for the discrete Gaussian component of the fluctuations of the height function in random tilings, as in Conjecture 24.6.*

Of course, both Claim 24.7 and Claim 24.8 describe the same object. The advantage of Claim 24.8 is that we are actually able to prove it in a forthcoming paper.[4] Nevertheless, the answers in these claims also look visually different. Hence, we are led to proving the following statement:

Proposition 24.9   *Quadratic forms (24.12) and (24.16) are the same.*

Proof   The key idea is to split the minimization procedure in (24.12) into two steps. First, we will *fix* the values of h on the real axis in an arbitrary way subject to (24.9), (24.10), and (24.11) and minimize over all such functions.

---

[3] Ibid.
[4] Ibid.

The minimizer is then a harmonic function separately in the upper half-plane and in the lower half-plane; by symmetry, the values differ by conjugation of the arguments. By the Schwarz integral formula (i.e., Poisson kernel), this harmonic function can be expressed as an integral over the real axis

$$h(z) = \Re \left[\frac{1}{\pi \mathbf{i}} \int_{-\infty}^{\infty} \frac{h(\zeta)}{\zeta - z} d\zeta \right], \quad z \in \mathbb{H}, \tag{24.17}$$

and by a similar expression with a changed sign in the lower half-plane $\bar{\mathbb{H}}$.

In the second step we minimize the Dirichlet energy over the choices of the values of h on the real axis. We claim that the latter optimization is the same as the log-energy optimization of (24.16). Let us explain this.

We identify a function on the real axis $h(x)$ with a signed measure ν through $h(y) - h(x) = \nu([x, y])$. Although, a priori, the functions do not have to be differentiable and the signed measures do not have to be absolutely continuous, the minimum in (24.12) and (24.16) is attained on differentiable functions and measures with densities, respectively. Thus, we can restrict ourselves to the latter, in which case the derivative $h_x(x)$ becomes the density of ν. Clearly, under such identification, conditions on h (24.9)–(24.11) become conditions on ν (24.14). It thus remains to show that the functional (24.12) turns into (24.16).

Let us compute the Dirichlet energy of $h(x + \mathbf{i}y)$ given by (24.17) in the domain $y \geq \varepsilon > 0$. Integrating by parts, we get

$$\iint_{y \geq \varepsilon} \left[(h_x)^2 + (h_y)^2 \right] dx dy$$

$$= \int_{\varepsilon}^{+\infty} dy \left[(hh_x)(+\infty + \mathbf{i}y) - (hh_x)(-\infty + \mathbf{i}y) - \int_{-\infty}^{\infty} hh_{xx} dx \right]$$

$$+ \int_{-\infty}^{\infty} dx \left[(hh_y)(x + \mathbf{i}\infty) - (hh_y)(x + \mathbf{i}\varepsilon) - \int_{\varepsilon}^{\infty} hh_{yy} dy \right]. \tag{24.18}$$

Further, note that because $h(x)$ has a compact support on the real axis, $h(z)$ given by (24.17) decays as $O(1/|z|)$ when $z \to \infty$, and its derivatives decay as $O(1/|z|^2)$. Hence, all the boundary terms at the infinity in (24.18) vanish. Moreover, because h is harmonic, $hh_{xx} + hh_{yy} = 0$, and the double integral vanishes as well. We conclude that (24.18) is

$$- \int_{-\infty}^{\infty} (hh_y)(x + \mathbf{i}\varepsilon) dx = - \int_{-\infty}^{\infty} dx \, h(x + \mathbf{i}\varepsilon) \Re \left[\frac{1}{\pi} \int_{-\infty}^{\infty} \frac{h(\zeta)}{-(\zeta - (x + \mathbf{i}\varepsilon))^2} d\zeta \right].$$

We integrate by parts both in x and in ζ to get

$$- \frac{1}{\pi} \int_{-\infty}^{\infty} dx \, h_x(x + \mathbf{i}\varepsilon) \Re \left[\int_{-\infty}^{\infty} h_x(\zeta) \ln(\zeta - (x + \mathbf{i}\varepsilon)) d\zeta \right].$$

Note that we need to choose some branch of the logarithm here, but it does not matter, because we will get the logarithm of the absolute value when computing the real part anyway.

At this point, it remains to send $\varepsilon \to 0$ (the integrand is singular, and we also have h_x rather than h itself; hence, the limit needs some justifications, which we omit) and to add the same contribution from the lower half-plane to get the desired

$$\iint_{\mathbb{D}} |\nabla h|^2 dx dy = -\frac{2}{\pi} \int_{-\infty}^{\infty} \int_{-\infty}^{\infty} h_x(x) h_x(\zeta) \ln |\zeta - x| \, dx d\zeta,$$

which matches (24.16).   □

Remark 24.10   Although we were only concentrating on the discrete component, essentially the same arguments explain the link between covariance structures for the entire field of fluctuations: Gaussian free field in the 2D picture and universal covariance of random matrix theory and log-gases on the 1D section. This covariance coincidence is also discussed (through other tools) in Bufetov and Gorin (2019), Borodin et al. (2017).

Remark 24.11   It is interesting to also try to do a similar identification for the shifts – m in (24.6). This involves computing the second-order expansion of the logarithm of the partition function; we need $O(N)$ rather than $O(N^2)$ terms. On the side of log-gases, there is a certain understanding of how this can be done (at least for *continuous* log-gases). On the other hand, we are not aware of approaches to the computation of this second term in the 2D setting of random tilings.

24.5 Discrete Component in Tilings on Riemann Surfaces

There is another natural situation in which the height function of lozenge tilings asymptotically develops a discrete Gaussian component.

Let us give an example by looking at lozenge tilings of a torus, as we did in Lectures 3, 4, and 6. The height function does not make sense on a torus as a single-valued function: we define the heights by local rules, and these rules typically result in a nontrivial increment as we loop around the torus and come back to the same point.

There are two closely related ways to deal with this difficulty. We could either split the height function into two components: an affine multivalued part (sometimes called the "instanton part") keeping track of the height change as we loop around the torus and a scalar single-valued part. Alternatively, we can define the height function as a 1-form so that it is not a function of the point but, rather, of a path on the torus and therefore is allowed to have different values on paths representing different homotopy classes. In both ways, one can single out the component that asymptotically converges to a discrete Gaussian random variable: in the first approach, this is the instanton part, and in the second approach, these are the values of the 1-form on closed loops.

More generally, a similar component can be singled out whenever we deal with tilings of domains embedded into Riemann surfaces of a higher genus. We can deal with random functions of homotopy classes of closed loops, and we can ask about the asymptotic of such functions as the mesh size goes to zero.

In several situations the convergence of the discrete component of the tilings on Riemann surfaces to discrete Gaussian random variables has been rigorously proved; see Boutillier and De Tilière (2009), Dubedat (2015), Kenyon et al. (2016), Dubedat and Gheissari (2015), and Berestycki et al. (2019).

Lecture 25: Sampling Random Tilings

In this lecture we are interested in algorithms for sampling a random lozenge tiling. Fix a tileable domain R. We consider the cases in which the probability of a particular tiling of R is either uniform or proportional to q^{Volume}, where Volume is the number of cubes that one needs to add to the minimal tiling of R in order to obtain a given tiling.

We are interested in reasonably fast algorithms for sampling a tiling. The trivial algorithm – generating the set of all tilings of R and picking a tiling from this set according to some probability measure – is in general very slow in the size of R. Indeed, we saw in previous lectures that for a domain of linear size L, the number of tilings is $\exp(L^2 \cdots)$ (cf. Theorem 5.15). Fortunately, there are other sampling algorithms that do not require generating a list of all tilings. We will discuss a few algorithms that are based on Markov chains.

One practical reason for wanting fast ways to sample tilings from a distribution sufficiently close to the desired distribution is so that one can guess the peculiar features of the tiling model ($L^{1/3}$ behavior, Tracy–Widom distributions, Gaussian free field, etc.) by looking at some sampled tilings.

25.1 Markov Chain Monte Carlo

We start from a general theorem; see, for example, Theorem 1 in Section 1.12 in Shiryaev (2016).

Theorem 25.1 *Take a (time-homogeneous discrete-time) Markov chain with the finite state space $[n]$ and the transition matrix P. That is, the matrix entry P_{ij} is the probability of transitioning from state i to state j in one time step. Suppose that there exists an integer k, such that for all pairs of states (i, j), $(P^k)_{ij} > 0$.*

Then P has a unique invariant distribution $\boldsymbol{\pi}$ (i.e., a distribution vector $\boldsymbol{\pi}$ for which $\boldsymbol{\pi}P = \boldsymbol{\pi}$). Furthermore, for any distribution vector $\boldsymbol{x}$,

$$\lim_{t \to \infty} \boldsymbol{x}P^t = \boldsymbol{\pi}.$$

This theorem motivates the following general idea for a sampling algorithm. Design a Markov chain such that the assumptions of the previous theorem are satisfied. Pick your favorite start state (the corresponding distribution vector $\boldsymbol{x}$ has a 1 in the corresponding location and 0's elsewhere), and perform random transitions on it for long enough (so as to make the distribution of the end state as close to uniform as desired). Output the end state.

All Markov chains we consider in this lecture are going to have a state space equal to the set of all tilings of our domain R. In each step, the Markov chain will either add a cube to the tiling or remove a cube from the tiling. For our Markov chains, we want the assumptions of Theorem 25.1 to be satisfied. In particular, we want any tiling to be reachable from any other tiling by a sequence of cube-addition or cube-removal moves. This condition is guaranteed to hold when the domain R is a simply connected region of the plane, but it might fail to hold for domains with holes because adding or removing cubes does not change the height function on the hole boundaries, whereas there are generally multiple possibilities for the height of a hole. However, for the fixed height functions of the holes, any tiling is reachable from any other by a sequence of such moves, and the rest of the discussion in this lecture applies to that case (with appropriate modifications). So for domains that are subsets of the plane that are not simply connected but for which the distribution of the height function on the holes is known, one can still sample a random tiling by first sampling the heights of the holes from the known distribution, then sampling a random tiling, given these heights for the holes. But for clarity, we will from now on only consider the case in which R is a *simply connected* region of the plane.

Consider the Markov chain given by the following step:

Chain 25.2 Choose a vertex m of R uniformly at random, and flip a coin. With probability $1/2$, add a cube at m (if possible; otherwise, do nothing), and with probability $1/2$, remove a cube at m (if possible; otherwise, do nothing). The possible moves are shown in Figure 25.1.

When running this chain in practice, one can, of course, restrict to only picking vertices m at which it is possible to either add or remove a cube. However, the previous formulation is what we will use formally because it has a number of beneficial properties. Note that Chain 25.2 satisfies the assumption

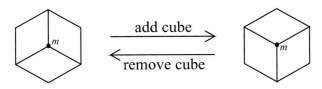

Figure 25.1 Adding/removing a cube at the vertex m.

of Theorem 25.1 because any tiling can be reached from any other, and there is always a nonzero probability of not changing the tiling, from which the assumption of the theorem follows. Hence, this chain has a unique invariant distribution, and we have the following proposition:

Proposition 25.3 *The uniform distribution is the invariant distribution of Chain 25.2.*

Proof For a general Markov chain, to show that π is an invariant distribution, it suffices to show reversibility; that is, $\pi_i P_{ij} = \pi_j P_{ji}$. Indeed, if this holds, then

$$(\pi P)_j = \sum_i \pi_i P_{ij} = \sum_i \pi_j P_{ji} = \pi_j \sum_i P_{ji} = \pi_j \implies \pi P = \pi.$$

So it remains to show that for our cube-switching Markov chain, π = the uniform distribution satisfies $\pi_i P_{ij} = \pi_j P_{ji}$. This is just a case check:

1. If $i = j$, then this is trivially true.
2. If tilings i and j are distinct but do not differ by adding/removing a single cube, then $P_{ij} = P_{ji} = 0$, so we are done.
3. If tilings i and j differ by adding/removing a single cube, then both P_{ij} and P_{ji} are $\frac{1}{2} \cdot \frac{1}{\text{number of vertices}}$, so we are done because for the uniform distribution, $\pi_i = \pi_j$. Note that it is crucial here that our Markov chain step involves picking a random vertex, regardless of whether a cube flip can be performed there.   □

As for the q^{Volume} distribution, consider the following Markov chain that is the same as Chain 25.2, except with different probabilities for adding/removing cubes:

Chain 25.4 Choose a vertex m of R uniformly at random, and flip a biased coin. With probability $\frac{q}{1+q}$, add a cube at m (if possible; otherwise, do nothing), and with probability $\frac{1}{1+q}$, remove a cube at m (if possible; otherwise, do nothing).

As argued before, Chain 25.4 also has a unique invariant distribution. We have the following proposition:

Proposition 25.5 *The q^{Volume} distribution is the invariant distribution of Chain 25.4.*

Proof As in the proof of Proposition 25.3, it suffices to show reversibility (i.e., that $\pi_i P_{ij} = \pi_j P_{ji}$). This is just a case check:

1. If $i = j$, then this is trivially true.
2. If tilings i and j are different but do not differ by adding/removing a single cube, then $P_{ij} = P_{ji} = 0$, so we are done.
3. If tilings i and j differ by adding/removing a single cube, with (without loss of generality) tiling j having one cube more than tiling i, then for the q^{Volume} distribution π, $\frac{\pi_j}{\pi_i} = q$, and with our transition probabilities, $\frac{P_{ij}}{P_{ji}} = \frac{\frac{q}{1+q}}{\frac{1}{1+q}} = q$. The desired equality $\pi_i P_{ij} = \pi_j P_{ji}$ follows. $\qquad\square$

Thus, we have the following algorithm for sampling a random tiling:

Algorithm 25.6 Pick your favorite tiling. Run either Chain 25.2 (for the uniform distribution) or Chain 25.4 (for the q^{Volume} distribution) long enough, and output the result.

Of course, this does not give exactly the distribution we are interested in for any finite time, but by Theorem 25.1, as the time goes to infinity, it gives an arbitrarily good approximation. One may ask what the runtime should be for the approximation to be ''sufficiently good.'' To be more rigorous, define the total variation distance between two distributions μ_1, μ_2 on a finite space X to be $\sum_{x \in X} |\mu_1(x) - \mu_2(x)|$. Define the mixing time of this Markov chain to be the minimal amount of time t the algorithm needs to run (from any initial state) so that for all $t' \geq t$, the variation distance between the induced distribution and the desired distribution is at most some fixed constant ϵ. In Randall and Tetali (2000), an upper bound $O(L^8 \ln L)$ for the mixing time in the domain of linear size L was obtained. For a slightly different Markov chain (that allows addition/removal of $1 \times 1 \times k$ stacks of cubes; it was introduced in Luby et al. (1995)), the mixing time was shown to be $O(L^4 \ln L)$ in Wilson (2004), and for hexagons, this bound is tight up to a constant factor. Section 5.6 in Wilson (2004) notes that the mixing time for this chain (adding only one cube at a time) is expected to be of the same order for "good" domains. This prediction, which also agrees with expectations from the theoretical physics literature, matches more recent results on the mixing time for this chain in specific classes of domains (Caputo et al., 2001; Laslier and Toninelli, 2015).

A general behavior (cf. Levin et al., 2017) is that before some time, the chain is not mixed at all, and after that time, it is mixed very well. To get an idea of what these estimates mean in practice, for instance, when our domain is a hexagon of side length 1000, one should run the Markov chain on a personal computer for about an entire day to get a good approximation.

25.2 Coupling from the Past (Propp and Wilson, 1996)

In any finite amount of time, the previous algorithm only gives a sample from an approximation of the desired distribution. This is often inconvenient because we would like to look at very delicate features of the tilings, and therefore, knowing that our sample is exact rather than approximate becomes important. With an upgrade known as "coupling from the past," we can sample from the desired distribution without any error.

Let us first discuss some ideas that will eventually lead us to the coupling from the past algorithm. Suppose we can run our Markov chain in a coupled way from all states, that is, picking a sequence of transitions (which we think of as transitions from any state via some canonical identification, not just transitions from some fixed state) and applying this same sequence to all initial states. Further suppose that at some time t, we notice that we have reached the same final state from all initial states. Suppose we output this final state. Because this state is independent of the initial configuration, one might expect that it is a sample from the invariant distribution because we can imagine starting from the invariant distribution originally. However, there are a few issues here.

First, with the algorithm as stated, we still run into our original problem – the total number of states is huge, so it would take a huge amount of time to simultaneously run this procedure from all start states. It turns out that this problem can be fixed because transitions preserve the height poset structure on tilings. This will be explained in more detail in the next few paragraphs.

The second issue is that there is no way to choose the desired time t deterministically, whereas if t is random (e.g., we choose it as the first time when all initial states lead to the same current state), then it becomes unclear whether the distribution of the output of such an algorithm is the invariant distribution of the Markov chain. Unfortunately, this turns out to be false in general, as the following counterexample shows. Suppose we have a Markov chain that satisfies the assumptions of Theorem 25.1 with more than one state and with a state into which one can transition from exactly one state (also counting the state itself). It is not hard to see that such Markov chains exist. Then, the probability of outputting this state at the *first* time t is clearly zero. However, under

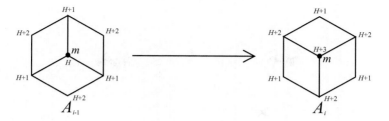

Figure 25.2 The configuration and height function around m in A_{i-1} and A_i.

the assumptions of Theorem 25.1, the invariant distribution cannot assign zero probability to some state. Thus, the distribution of the output of our algorithm is not the invariant distribution.

Let us first explain how to work around the issue of not being able to run the Markov chain from all initial states.

Proposition 25.7 *Suppose we have two tilings A_0 and B_0, such that at each grid vertex m, the height of A_0 is at most the height of B_0. Suppose we have a sequence of transitions $T_1, T_2, \ldots, T_t$. That is, each transition is a specification of a vertex m at which to perform an operation and whether this operation is adding or removing a cube. Let A_t and B_t be the tilings we obtain after performing this sequence of operations on A_0 and B_0, respectively. Then, at every grid point m, the height of A_t is at most the height of B_t.*

Proof For a contradiction, say that there is a vertex at which the height of A_t is greater than the height of B_t. Then there must be a first transition after which there exists such a vertex; let it be T_i. This means that at all grid vertices, the height of the tiling A_{i-1} was at most that of the tiling B_{i-1}, but there exists a vertex m at which the height of A_i is greater than the height of B_i. Our operations only change the height at one vertex – the vertex at which we add or remove a cube. So in our situation, we must have added or removed a cube at the vertex m, and at all neighboring vertices, the height of A_i is still at most the height of B_i. Let us consider the case in which we added a cube to A_{i-1} at m. The case of removing a cube from B_{i-1} at m is completely analogous. Because the height of A_i became greater than that of B_i at m, we must have not added a cube to B_i, so $B_i = B_{i-1}$. Because we know the local configuration of the tilings A_{i-1} and A_i around the vertex m and that the heights of other vertices are preserved, we can write out the height functions in terms of the height H of m in A_{i-1}. These are shown in Figure 25.2.

Note that in B_{i-1}, m has a height of at most $H+2$ by our assumption about the height function of A_i becoming larger than that of B_i. Also note that the vertices

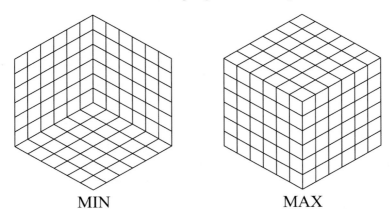

MIN   MAX

Figure 25.3 MIN and MAX tilings for the hexagon.

adjacent to m in the three negative coordinate directions have heights of at least $H + 2$ in B_{i-1} (see Figure 25.2), so, in particular, they have heights greater than the height of m. Hence, the edges between m and these three vertices must be diagonals of lozenges in B_{i-1}. But this implies that we could have also added a cube to B_{i-1}, a contradiction. $\square$

Recall from the first lecture that for any simply connected domain R, there is a minimal tiling (for the hexagon, this is the hollowed-out cube, as in Figure 25.3) which has height function at most that of any other tiling at all points. Also notice that there is a maximal tiling (for the hexagon, this is the filled-in cube, as in Figure 25.3) that has a height function that is at least that of any other tiling at all points. Let us call these tilings MIN and MAX. By Proposition 25.7, if a sequence of transitions leads to the same tiling X from both MIN and MAX, then this sequence must lead to X from all initial tilings because the height function of any final state is both lower-bounded and upper-bounded by the height function of X at all points. Thus, this solves the issue of having to run the Markov chain from all states – now we just need to run it from MIN and MAX to be able to tell when all branches coalesce.

We still have the problem that the previous algorithm of applying successive transitions might give the wrong distribution because of the randomness of t. It turns out that this is not the case if, instead of applying successive transitions to the end state, one applies all transitions in the opposite order (i.e., starting from the transition that was chosen last). We consider the following algorithm:

Algorithm 25.8 (Coupling from the Past) Successively generate random transitions $T_{-1}, T_{-2}, \ldots$. That is, each transition is a specification of a vertex

m at which to perform an operation and whether this operation is adding or removing a cube, generated at random with the same probability distribution as before (in either Chain 25.2 or Chain 25.4). After generating each new transition T_{-t}, apply the same sequence of transitions $T_{-t} \circ T_{-(t-1)} \circ \cdots \circ T_{-1}$ to both MIN and MAX. If the two tilings obtained from both start states are the same, output this tiling (and halt).

In practice, it is more efficient to only apply the transitions to MIN and MAX after generating $t = 1, 2, 4, 8, \ldots$ from them. It is easy to see that this algorithm halts almost surely. Let us now prove that this algorithm generates samples from exactly the desired distribution.

Proposition 25.9 *The distribution of the output of Algorithm 25.8 is precisely the invariant distribution of the corresponding tiling Markov chain (uniform or q^{Volume}).*

Proof If $T_{-t} \circ T_{-(t-1)} \circ \cdots \circ T_{-1}$ takes both MIN and MAX to the same end state i, then it also does so for any other state (by Proposition 25.7). The crucial observation is that if $T_{-t} \circ T_{-(t-1)} \circ \cdots \circ T_{-1}$ takes all initial states to i, then for $t' > t$, $T_{-t'} \circ T_{-(t'-1)} \circ \cdots \circ T_{-1} = (T_{-t'} \circ T_{-(t'-1)} \circ \cdots \circ T_{-(t+1)}) \circ (T_{-t} \circ T_{-(t'-1)} \circ \cdots \circ T_{-1})$ also takes any initial state to i. Starting from any start state (i.e., a distribution vector $\boldsymbol{x}$ with a 1 in the corresponding location and 0's elsewhere), Theorem 25.1 gives that $\boldsymbol{x}P^t$ approaches the invariant distribution as $t \to \infty$. Because for any particular sequence of transitions $\ldots, T_{-2}, T_{-1}$, there almost surely exists a time t after which the end state is unchanged by adding transitions to the beginning, the limiting distribution (formally, the limiting distribution starting from some fixed start state, although it does not matter because the result is the same for all start states) is also equal to the distribution of the end states generated by our algorithm. Hence, the distribution of these end states is equal to the invariant distribution of the Markov chain. This is what we wanted to show. □

To recap, one advantage Algorithm 25.8 has is that it allows sampling from the exact distribution we desire, whereas the simple Markov chain Monte Carlo algorithm considered before only samples from a good approximation of this distribution. The runtime is unbounded, but the algorithm almost surely stops in finite time. For the hexagon, the expected runtime is of the same order as the mixing time. Perfect sampling of random lozenge tilings was one of the important initial motivations for developing coupling from the past in Propp and Wilson (1996), yet it is also useful for sampling from other large systems (e.g., the six-vertex model).

Although coupling from the past is a powerful and simple to implement method, it is by no means a unique way to sample random tilings. In the next sections we briefly mention other approaches.

25.3 Sampling through Counting

Jerrum et al. (1986) popularized a powerful idea: counting and sampling are closely related to each other. In our context, imagine that we could quickly compute the probability that a given position in the domain is occupied by the lozenge of a fixed type. Then we could flip a three-sided coin to choose between $\diamond$, $\varnothing$, and $\varnothing$ for this position, and continue afterwards for the smaller domain.

We learned in Lectures 2 and 3 that the required probabilities can be computed either by inverting the Kasteleyn matrix or through evaluation of the determinants of matrices built out of the binomial coefficients. Thus, the sampling now reduces to the well-studied linear algebra problems of fast computations for the determinants and inverse matrices. These ideas were turned into sampling algorithms in Colbourn et al. (1996) for spanning trees and in Wilson (1997) for tilings.

A fast recursive algorithm for computing the same probabilities that works for very general weights is further discussed in Propp (2003).

25.4 Sampling through Bijections

Here is another idea. In many situations the total number of tilings (or weighted sum over tilings) is given by a simple product formula, as in Theorem 1.1 of Lecture 1 or (22.9) in Lecture 22. If we can turn the enumeration identity into a bijection between tilings and some simpler objects enumerated by the same product, then as soon as we manage to sample these objects, we get access to random tilings as well.

Perhaps the most famous of such bijections is the Robinson–Schensted–Knuth (RSK) correspondence between rectangular matrices filled with integers and pairs of semistandard Young tableaux; for example, see Section 5.3 in Romik (2015) and Section 10.2 in Baik et al. (2016) for reviews in a probabilistic context, and see Sagan (2001) for connections to representation theory and symmetric functions. It is straightforward to sample a rectangular matrix with independent matrix elements, and using RSK, we immediately get random pairs of Young tableaux, which, in turn, can be identified with plane partitions. This approach can be used for the sampling of q^{Volume}-weighted plane partitions of Lecture 22. For the Schur processes, which generalize plane partitions, the detailed exposition of the approach can be found in Betea et al. (2018).

More delicate bijections were used in Bodini et al. (2010) for sampling random plane partitions with a fixed volume and in Krattenthaler (1999b) for sampling the random lozenge tilings of hexagons.

25.5 Sampling through Transformations of Domains

There are specific domains for which the number of tilings is small and sampling is easy. This could be either because the domain itself is very small or because the height function is extremal along the boundary. For instance, the $A \times B \times 0$ hexagon has a unique tiling.

A central idea for a class of algorithms is to introduce a Markov chain (with local transition rules) that takes a uniformly random tiling of a domain as an input and gives a uniformly random tiling of a slightly more complicated domain as an output. In this way, we can start from a domain with a single tiling and reach much more complicated regions in several steps. The main difficulty in this approach is to design such a Markov chain because there is no universal recipe to produce it.

The first example of such a chain is known as the "shuffling algorithm" for the Aztec diamond (Elkies et al., 1992). It works with tilings of rhombuses drawn on the square grid ("diamonds") with 2×1 dominoes, as in Figure 1.10 of Lecture 1. Starting from a 2×2 square, which has two tilings, it grows the size of the domain, ultimately getting to huge rhombuses.

There are several different points of view on the shuffling algorithm, which lead to different generalizations. The first point of view is that it provides a stochastic version of a deterministic operation that modifies a bipartite graph through a sequence of local moves (known as "urban renewal" or "spider move") with an explicitly controlled change of the (weighted) count of all perfect matchings. It is explained in Propp (2003) how this can be used for sampling random domino tilings with quite generic weights. Because the admissible weights in this procedure are essentially arbitrary, it is very flexible: by various degenerations, one can get all kinds of complicated domains, tilings on other lattices (including lozenge tilings), and so forth. Extending beyond the probabilistic context, urban renewal is studied as an abstract algebraic operation on weighted bipartite graphs in Goncharov and Kenyon (2013); it was further linked to the Miquel move on cycle patterns (arrangements of intersecting circles on the plane) in Affolter (2018) and Kenyon et al. (2018).

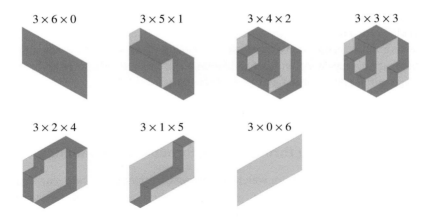

Figure 25.4 Steps of the shuffling algorithm for lozenge tilings of hexagons.

Another point of view exploits the connections of the domino tilings to Schur polynomials (which is similar to the lozenge tilings case we saw in Proposition 19.3 and Lecture 22; see also Bufetov and Knizel (2018)). Along these lines, Borodin (2011a) constructed an efficient sampler for q^{Volume}-weighted plane partitions, which we discussed in Lectures 22 and 23, and their skew versions.

The third point of view treats domino tilings as orthogonal polynomial ensembles; see the footnote to Exercise 21.7, and see Cohn et al. (1996) and Johansson (2005). Along these lines, Borodin and Gorin (2009) and Borodin et al. (2010) designed a Markov chain on the random lozenge tilings of a hexagon that changes the side lengths, thus allowing them to reach an $A \times B \times C$ hexagon starting from the degenerate $A \times (B + C) \times 0$ hexagon; see Figure 25.4.

The aforementioned points of view and generalizations of the shuffling algorithm share one feature: if we restrict our attention to a subset of tiles, then the time evolution turns into a one-dimensional interacting particle system, which is a suitable discrete version of the totally asymmetric simple exclusion process (TASEP). The TASEP itself is a prominent continuous-time dynamic on particle configurations on the integer lattice $\mathbb{Z}$: each particle has an independent exponential clock (Poisson point process), and whenever the clock rings, the particle jumps one step to the right unless that spot is occupied ("exclusion"); then the clock is restarted, and the process continues in the same way.

The link between the shuffling algorithm for the domino tilings of the Aztec diamond and a discrete-time TASEP was noticed and exploited for the identification of the boundary of the frozen region in Jockusch et al. (1995). For lozenge tilings (of unbounded domains), a direct link to the continuous-time TASEP was developed in Borodin and Ferrari (2014). These results connect the edge-limit theorems for tilings of Lecture 18 to the appearance of the $t^{1/3}$ scaling and Tracy–Widom distribution in the large-time asymptotic of TASEP.[1]

The connection between statistical mechanics models (such as random tilings) and interacting particle systems is a fruitful topic, and we refer the reader to Borodin and Gorin (2016) and Borodin and Petrov (2014, 2017) for further reviews.

---

[1] The Tracy–Widom fluctuations for TASEP were first established by Johansson (2000).

References

Adler, M., Johansson, K., and van Moerbeke, P. (2018a). Lozenge tilings of hexagons with cuts and asymptotic fluctuations: a new universality class. *Mathematical Physics, Analysis and Geometry*, **21**, 1–53. arXiv:1706.01055.

Adler, M., Johansson, K., and van Moerbeke, P. (2018b). Tilings of nonconvex polygons, skew-Young tableaux and determinantal processes. *Communications in Mathematical Physics*, **364**, 287–342. arXiv:1609.06995.

Affolter, N. C. (2018). Miquel dynamics, Clifford lattices and the dimer model [Preprint]. arXiv:1808.04227.

Aggarwal, A. (2015). Correlation functions of the Schur process through Macdonald difference operators. *Journal of Combinatorial Theory, Series A*, **131**, 88–118. arXiv:1401.6979.

Aggarwal, A. (2019). Universality for lozenge tilings local statistics [Preprint]. arXiv:1907.09991.

Aggarwal, A., and V. Gorin (2021). Gaussian Unitary Ensemble in random lozenge tilings [Preprint]. arXiv:2106.07589.

Ahn, A. (2018). Global universality of Macdonald plane partitions. *Annales Institut Henri Poincaré Probability and Statistics*, **56**(3), 1641–1705. arXiv:1809.02698.

Aissen, M., Schoenberg, I. J., and Whitney, A. (1952). On the generating functions of totally positive sequences I. *Journal d'Analyse Mathématique*, **2**, 93–103.

Akemann, G., Baik, J., and Di Francesco, P., eds. (2011). *The Oxford Handbook of Random Matrix Theory*. Oxford University Press.

Anderson, G. W., Guionnet, A., and Zeitouni, O. (2010). *An Introduction to Random Matrices*. Cambridge University Press.

Ardila, F., and Stanley, R. P. (2010). Tilings. *The Mathematical Intelligencer*, **32**, 32–43. arXiv:math/0501170.

Astala, E. K., Duse, E., Prause, I., and Zhong, X. (2020). Dimer models and conformal structures [Preprint]. arXiv:2004.02599.

Azuma, K. (1967). Weighted sums of certain dependent random variables. *Tohoku Mathematical Journal, Second Series*, **19**(3), 357–367.

Baik, J., Deift, P., and Suidan, T. (2016). *Combinatorics and Random Matrix Theory*. Graduate Studies in Mathematics, vol. 172. American Mathematical Society.

Baik, J., Kriecherbauer, T., McLaughlin, K. T. R., and Miller, P. D. (2003). Uniform asymptotics for polynomials orthogonal with respect to a general class of discrete weights and universality results for associated ensembles. *International Mathematics Research Notices*, **15**. arXiv:math/0310278.

Baryshnikov, Y. (2001). GUEs and queues. *Probability Theory and Related Fields*, **119**(2), 256–274.

Basok, M., and Chelkak, D. (2018). Tau-functions à la Dubédat and probabilities of cylindrical events for double-dimers and CLE(4) [to appear in *Journal of European Mathematical Society*]. arXiv:1809.00690.

Baxter, R. (2007). *Exactly Solved Models in Statistical Mechanics*. Dover.

Beffara, V., Chhita, S., and Johansson, K. (2020). Local geometry of the rough-smooth interface in the two-periodic Aztec diamond [Preprint]. arXiv:2004.14068.

Berestycki, N., Laslier, B., and Ray, G. (2019). The dimer model on Riemann surfaces, I [Preprint]. arXiv:1908.00832.

Berestycki, N., Laslier, B., and Ray, G. (2020). Dimers and imaginary geometry. *Annals of Probability*, **48**(1), 1–52. arXiv:1603.09740.

Berggren, T., and Duits, M. (2019). Correlation functions for determinantal processes defined by infinite block Toeplitz minors. *Advances in Mathematics*, **356**(2019), paper 106766. arXiv:1901.10877.

Betea, D., Boutillier, C., Bouttier, J., Chapuy, G., Corteel, S., and Vuletić, M. (2018). Perfect sampling algorithm for Schur processes. *Markov Processes and Related Fields*, **24**, 381–418. arXiv:1407.3764

Blöte, H. W. J., and Hilhorst, H. J. (1982). Roughening transitions and the zero-temperature triangular Ising antiferromagnet. *Journal of Physics A: Mathematical and General*, **15**, 631–637.

Bodineau, T., Schonmann, R. H., and Shlosman, S. (2005). 3D crystal: how flat its flat facets are? *Communications in Mathematical Physics*, **255**, 747–766. arXiv:math-ph/0401010.

Bodini, O., Fusy, E., and Pivoteau, C. (2010). Random sampling of plane partitions. *Combinatorics, Probability and Computing*, **19**(2), 201–226. arXiv:0712.0111.

Borodin, A. (2011a). Determinantal point processes. In G. Akermann, J. Baik, P. Di Franceco, eds., *Oxford Handbook of Random Matrix Theory*. Oxford University Press.

Borodin, A. (2011b). Schur dynamics of the Schur processes. *Advances in Mathematics*, **228**(4), 2268–2291. arXiv:1001.3442.

Borodin, A., and Corwin, I. (2014). Macdonald processes. *Probability Theory and Related Fields*, **158**(1–2), 225–400. arXiv:1111.4408.

Borodin, A., Corwin, I., Gorin,V., and Shakirov, S. (2016). Observables of Macdonald processes. *Transactions of American Mathematical Society*, **368**, 1517–1558. arxiv:1306.0659.

Borodin, A., and Ferrari, P. (2014). Anisotropic growth of random surfaces in 2 + 1 dimensions. *Communications in Mathematical Physics*, **325**(2), 603–684. arXiv:0804.3035.

Borodin, A., and Gorin, V. (2009). Shuffling algorithm for boxed plane partitions. *Advances in Mathematics*, **220**(6), 1739–1770. arXiv:0804.3071.

Borodin, A., and Gorin, V. (2015). General beta Jacobi corners process and the Gaussian free field, *Communications on Pure and Applied Mathematics*, **68**(10), 1774–1844. arXiv:1305.3627.

Borodin, A., and Gorin, V. (2016). Lectures on integrable probability. In V. Sidoravicius and S. Smirnov, eds., *Probability and Statistical Physics in St. Petersburg*,

Proceedings of Symposia in Pure Mathematics, vol. 91. American Mathematical Society. arXiv:1212.3351.

Borodin, A., Gorin, V., and Guionnet, A. (2017). Gaussian asymptotics of discrete β-ensembles. *Publications mathématiques de l'IHÉS*, **125**(1), 1–78. arXiv:1505.03760

Borodin, A., Gorin, V., and Rains, E. M. (2010). q-Distributions on boxed plane partitions. *Selecta Mathematica*, **16**(4), 731–789. arXiv:0905.0679

Borodin, A., and Kuan, J. (2008). Asymptotics of Plancherel measures for the infinite-dimensional unitary group. *Advances in Mathematics*, **219**(3), 894–931. arXiv:0712.1848

Borodin, A., and Kuan, J. (2010). Random surface growth with a wall and Plancherel measures for O(infinity). *Communications on Pure and Applied Mathematics*, **63**(7), 831–894. arXiv:0904.2607

Borodin, A., and Olshanski, G. (2012). The boundary of the Gelfand–Tsetlin graph: a new approach. *Advances in Mathematics*, **230**, 1738–1779; arXiv:1109.1412.

Borodin, A., and Olshanski, G. (2013). The Young bouquet and its boundary. *Moscow Mathematical Journal*, **13**(2), 193–232. arXiv:1110.4458.

Borodin, A., and Olshanski, G. (2016). *Representations of the Infinite Symmetric Group*. Cambridge University Press.

Borodin, A., and Petrov, L. (2014). Integrable probability: from representation theory to Macdonald processes. *Probability Surveys*, **11**, 1–58. arXiv:1310.8007.

Borodin, A., and Petrov, L. (2017). Lectures on integrable probability: stochastic vertex models and symmetric functions. In G. Schehr, A. Altland, Y. V. Fyodorov, N. O'Connell, and L. F. Cugliandolo, eds., *Stochastic Processes and Random Matrices: Lecture Notes of the Les Houches Summer School, July 2015*, vol. 104. Oxford University Press. arXiv:1605.01349.

Borodin, A., and Shlosman, S. (2010). Gibbs ensembles of nonintersecting paths. *Communications in Mathematical Physics*, **293**, 145–170. arXiv:0804.0564.

Borodin, A., and Toninelli, F. (2018). Two-dimensional anisotropic KPZ growth and limit shapes. *Journal of Statistical Mechanics*, (2018), 083205. arXiv:1806.10467.

Borot, G., and Guionnet, A. (2013). Asymptotic expansion of β matrix models in the one-cut regime. *Communications in Mathematical Physics*, **317**, 447. arXiv:1107.1167.

Boutillier, C., and De Tilière, B. (2009). Loop statistics in the toroidal honeycomb dimer model. *The Annals of Probability*, **37**(5), 1747–1777. arXiv:math/0608600.

Breuer, J., and Duits, M. (2017). Central limit theorems for biorthogonal ensembles and asymptotics of recurrence coefficients. *Journal of the American Mathematical Society*, **30**, 27–66. arXiv:1309.6224.

Brézin, E., and Hikami, S. (1996). Correlations of nearby levels induced by a random potential. *Nuclear Physics B*, **479**(3), 697–706. arXiv:cond-mat/9605046

Brézin, E., and Hikami, S. (1997). Spectral form factor in a random matrix theory. *Physical Review E*, **55**(4), 4067. arXiv:cond-mat/9608116.

Bufetov, A., and Gorin, V. (2015). Representations of classical Lie groups and quantized free convolution. *Geometric and Functional Analysis (GAFA)*, **25**(3), 763–814. arXiv:1311.5780.

Bufetov, A., and Gorin, V. (2018). Fluctuations of particle systems determined by Schur generating functions. *Advances in Mathematics*, **338**(7), 702–781. arXiv:1604.01110.

Bufetov, A., and Gorin, V. (2019). Fourier transform on high-dimensional unitary groups with applications to random tilings. *Duke Mathematical Journal*, **168**(13), 2559–2649. arXiv:1712.09925.

Bufetov, A., and Knizel, A. (2018). Asymptotics of random domino tilings of rectangular Aztec diamonds. *Annales Institut Henri Poincaré: Probability and Statistics*, **54**(3), 1250–1290. arXiv:1604.01491.

Bykhovskaya, A., and Gorin, V. (2020). Cointegration in large VARs [Preprint]. arXiv:2006.14179.

Caputo, P., Martinelli, F., and Toninelli, F. L. (2012). Mixing times of monotone surfaces and SOS interfaces: a mean curvature approach. *Communications in Mathematical Physics*, **311**, 157–189. arXiv:1101.4190.

Cerf, R., and Kenyon, R. (2001). The low-temperature expansion of the Wulff crystal in the 3D Ising model. *Communications in Mathematical Physics*, **222**(1), 147–179.

Charlier, C. (2020). Doubly periodic lozenge tilings of a hexagon and matrix valued orthogonal polynomials. *Studies in Applied Mathematics*, **146**(1), 3–80. arXiv:2001.11095.

Charlier, C., Duits, M., Kuijlaars, A. B. J., and Lenells, J. (2020). A periodic hexagon tiling model and non-Hermitian orthogonal polynomials. *Communications in Mathematical Physics*, **378**, 401–466. arXiv:1907.02460.

Chelkak, D., Laslier, B., and Russkikh, M. (2020). Dimer model and holomorphic functions on t-embeddings of planar graphs [Preprint]. arXiv:2001.11871.

Chelkak, D., and Ramassamy, S. (2020). Fluctuations in the Aztec diamonds via a Lorentz-minimal surface [Preprint]. arXiv:2002.07540.

Chhita, S., and Johansson, K. (2016). Domino statistics of the two-periodic Aztec diamond. *Advances in Mathematics*, **294**, 37–149. arXiv:1410.2385.

Chhita, S., and Young, B. (2014). Coupling functions for domino tilings of Aztec diamonds. *Advances in Mathematics*, **259**, 173–251. arXiv:1302.0615.

Cimasoni, D., and Reshetikhin, N. (2007). Dimers on surface graphs and spin structures. I. *Communications in Mathematical Physics*, **275**, 187–208. arXiv:math-ph/0608070.

Ciucu, M. (1997). Enumeration of perfect matchings in graphs with reflective symmetry. *Journal of Combinatorial Theory, Series A*, **77**(1), 67–97.

Ciucu, M. (2008). Dimer packings with gaps and electrostatics. *Proceedings of the National Academy of Sciences of the United States of America*, **105**(8), 2766–2772.

Ciucu, M. (2015). A generalization of Kuo condensation. *Journal of Combinatorial Theory, Series A*, **134**, 221–241. arXiv:1404.5003.

Ciucu, M. (2018). Symmetries of shamrocks, Part I. *Journal of Combinatorial Theory, Series A*, **155**, 376–397.

Ciucu, M., and Fischer, I. (2015). Proof of two conjectures of Ciucu and Krattenthaler on the enumeration of lozenge tilings of hexagons with cut off corners. *Journal of Combinatorial Theory, Series A*, **133**, 228–250. arXiv:1309.4640.

Ciucu, M., and Krattenthaler, C. (2002). Enumeration of lozenge tilings of hexagons with cut-off corners. *Journal of Combinatorial Theory, Series A*, **100**(2), 201–231. arXiv:math/0104058.

Cohn, H., Elkies, N., and Propp, J. (1996). Local statistics for random domino tilings of the Aztec diamond. *Duke Mathematical Journal*, **85**, 117–166.

Cohn, H., Larsen, M., and Propp, J. (1998). The shape of a typical boxed plane partition. *New York Journal of Mathematics*, **4**, 137–165. arXiv:math/9801059.

Cohn, J., Kenyon, R., and Propp, J. (2001). A variational principle for domino tilings. *Journal of the American Mathematical Society*, **14**(2), 297–346. arXiv:math/0008220.

Colbourn, C. J., Myrvold, W. J., and Neufeld, E. (1996). Two algorithms for unranking arborescences. *Journal of Algorithms*, **20**, 268–281.

Corwin, I. (2012). The Kardar–Parisi–Zhang equation and universality class. *Random Matrices: Theory and Applications*, **1**(1). arXiv:1106.1596.

Corwin, I., and Hammond, A. (2014). Brownian Gibbs property for Airy line ensembles. *Inventiones Mathematicae*, **195**(2), 441–508. arXiv:1108.2291.

Corwin, I., and Sun, X. (2014). Ergodicity of the Airy line ensemble. *Electronic Communications in Probability*, **19**, article 49. arXiv:1405.0464.

Daley, D. J., and Vere-Jones, D. (2003). *An Introduction to the Theory of Point Processes*, vol. 1, 2nd ed. Springer.

De Silva, D., and Savin, O. (2010). Minimizers of convex functionals arising in random surfaces. *Duke Mathematical Journal*, **151**(3), 487–532. arXiv:0809.3816.

Deift, P. (2000). *Orthogonal Polynomials and Random Matrices: A Riemann-Hilbert Approach*. American Mathematical Society.

Destainville, N., Mosseri, R., and Bailly, F. (1997). Configurational entropy of co-dimension one tilings and directed membranes. *Journal of Statistical Physics*, **87**, 697–754.

Dimitrov, E. (2020). Six-vertex models and the GUE-corners process. *International Mathematics Research Notices*, **6**, 1794–1881. arXiv:1610.06893.

Dubedat, J. (2009). SLE and the free field: partition functions and couplings. *Journal of the American Mathematical Society*, **22**(4), 995–1054. arXiv:0712.3018.

Dubedat, J. (2015). Dimers and families of Cauchy Riemann operators I. *Journal of the American Mathematical Society*, **28**(4), 1063–1167. arXiv:1110.2808.

Dubedat, J. (2019). Double dimers, conformal loop ensembles and isomonodromic deformations. *Journal of the European Mathematical Society*, **21**(1), 1–54. arXiv:1403.6076.

Dubedat, J., and Gheissari, R. (2015). Asymptotics of height change on toroidal Temperleyan dimer models. *Journal of Statistical Physics*, **159**(1), 75–100. arXiv:1407.6227.

Duits, M. (2018). On global fluctuations for non-colliding processes. *Annals of Probability*, **46**(3), 1279–1350. arXiv:1510.08248.

Duits, M., and Kuijlaars, A. B. J. (2019). The two periodic Aztec diamond and matrix orthogonal polynomials [to appear in *Journal of European Mathematical Society*]. arXiv:1712.05636.

Duse, E., and Metcalfe, A. (2015). Asymptotic geometry of discrete interlaced patterns: Part I. *International Journal of Mathematics*, **26**(11), 1550093. arXiv:1412.6653.

Duse, E., and Metcalfe, A. (2018). Universal edge fluctuations of discrete interlaced particle systems. *Annales Mathématiques Blaise Pascal*, **25**(1), 75–197. arXiv:1701.08535.

Duse, K. E., Johansson, K., and Metcalfe, A. (2016). The Cusp–Airy process. *Electronic Journal of Probability*, **21**, article 57. arXiv:1510.02057.

Edrei, A. (1953). On the generating function of a doubly infinite, totally positive sequence. *Transactions of American Mathematical Society*, **74**, 367–383.

Elkies, N., Kuperberg, G., Larsen, M., and Propp, J. (1992). Alternating-sign matrices and domino tilings. *Journal of Algebraic Combinatorics*, **1**(2), 111–132. arXiv:math/9201305.

Erdos, L., and Yau, H-T. (2017). *A Dynamical Approach to Random Matrix Theory*. Vol. 28 of Courant Lecture Notes in Mathematics. Courant Institute of Mathematical Sciences, New York; and American Mathematical Society.

Eynard, B., and Mehta, M. L. (1998). Matrices coupled in a chain. I. Eigenvalue correlations. *Journal of Physics A: Mathematical and General*, **31**, 4449–4456. arXiv:cond-mat/9710230.

Ferrari, P. L., and Spohn, H. (2003). Step fluctuations for a faceted crystal. *Journal of Statistical Physics*, **113**, 1–46. arXiv:cond-mat/0212456.

Forrester, P. J. (2010). *Log-Gases and Random Matrices*. Princeton University Press.

Gelfand, I. M., and Naimark, M. A. (1950). Unitary representations of the classical groups. *Trudy Matematicheskogo Instituta imeni V. A. Steklova*, **36**, 3–288.

Gelfand, I. M., and Tsetlin, M. L. (1950). Finite-dimensional representations of the group of unimodular matrices. *Proceedings of the USSR Academy of Sciences*, **71**, 825–828.

Gessel, I., and Viennot, G. (1985). Binomial determinants, paths, and hook length formulae. *Advances in Mathematics*, **58**(3), 300–321.

Gilmore, T. (2017). Holey matrimony: marrying two approaches to a tiling problem. *Séminaire Lotharingien de Combinatoire*, **78B**, article 26.

Golomb, S. W. (1995). *Polyominoes: Puzzles, Patterns, Problems, and Packings*, 2nd ed. Princeton University Press.

Goncharov, A. B., and Kenyon, R. (2013). Dimers and cluster integrable systems. *Annales scientifiques de l'École Normale Supérieure*, **46**(5), 747–813. arXiv:1107.5588.

Gorin, V. (2008). Non-intersecting paths and Hahn orthogonal polynomial ensemble. *Functional Analysis and Its Applications*, **42**(3), 180–197. arXiv: 0708.2349.

Gorin, V. (2012). The q-Gelfand-Tsetlin graph, Gibbs measures and q-Toeplitz matrices. *Advances in Mathematics*, **229**(1), 201–266. arXiv:1011.1769.

Gorin, V. (2014). From alternating sign matrices to the Gaussian unitary ensemble. *Communications in Mathematical Physics*, **332**(1), 437–447, arXiv:1306.6347.

Gorin, V. (2017). Bulk universality for random lozenge tilings near straight boundaries and for tensor products. *Communications in Mathematical Physics*, **354**(1), 317–344. arXiv:1603.02707.

Gorin, V., and Olshanski, G. (2016). A quantization of the harmonic analysis on the infinite-dimensional unitary group. *Journal of Functional Analysis*, **270**(1), 375–418. arXiv:1504.06832.

Gorin, V., and Panova, G. (2015). Asymptotics of symmetric polynomials with applications to statistical mechanics and representation theory. *Annals of Probability*, **43**(6), 3052–3132. arXiv:1301.0634.

Gorin, V., and Zhang, L. (2018). Interlacing adjacent levels of β-Jacobi corners processes. *Probability Theory and Related Fields*, **172**(3–4), 915–981. arXiv:1612.02321.

Grünbaum, B., and Shephard, G. (2016). *Tilings and Patterns*, 2nd ed. Dover Books on Mathematics.

Guionnet, A. (2004). First order asymptotics of matrix integrals; a rigorous approach towards the understanding of matrix models. *Communications in Mathematical Physics*, **244**, 527–569. arXiv:math/0211131.

Guionnet, A. (2019). *Asymptotics of Random Matrices and Related Models: The Uses of Dyson–Schwinger Equations* (CBMS Regional Conference Series in Mathematics). American Mathematical Society.

Höffe, M. (1997). Zufallsparkettierungen und Dimermodelle. Unpublished Ph.D. thesis, Institut für Theoretische Physik der Eberhard-Karls-Universität, Tübingen.

Hu, X., Miller, L., and Peres, Y. (2010). Thick points of the Gaussian free field. *Annals of Probability*, **38**(2), 896–926. arXiv:0902.3842.

Izergin, A. G. (1987). Partition function of the six-vertex model in a finite volume. *Soviet Physics Doklady*, **32**, 878–879.

Jerrum, M. R., Valiant, L. G., and Vazirani, V. V. (1986). Random generation of combinatorial structures from a uniform distribution. *Theoretical Computer Science*, **43**, 169–188.

Jockusch, W., Propp, J., and Shor, P. (1995). Random domino tilings and the arctic circle theorem [Preprint]. arXiv:math/9801068.

Johansson, K. (1998). On fluctuations of eigenvalues of random Hermitian matrices. *Duke Mathematical Journal*, **91**(1), 151–204.

Johansson, K. (2000). Shape fluctuations and random matrices. *Communications in Mathematical Physics*, **209**, 437–476. arXiv:math/9903134.

Johansson, K. (2002). Non-intersecting paths, random tilings and random matrices. *Probability Theory and Related Fields*, **123**, 225–280. arXiv:math/0011250.

Johansson, K. (2005). The arctic circle boundary and the Airy process. *Annals of Probability*, **33**, 1–30. arXiv:math/0306216.

Johansson, K. (2006). Random matrices and determinantal processes. In A. Bovier, F. Dunlop, A. Van Enter, F. Den Hollander, and J. Dalibard, eds., *Mathematical Statistical Physics*. Elsevier. arXiv:math-ph/0510038.

Johansson, K. (2016). Edge fluctuations of limit shapes. *Current Developments in Mathematics 2016*. **2016**, 47–110. arXiv:1704.06035.

Johansson, K., and E. Nordenstam (2006). Eigenvalues of GUE minors, *Electronic Journal of Probability*, 11, paper no. 50, 1342–1371.

Johnstone, I. (2008). Multivariate analysis and Jacobi ensembles: largest eigenvalue, Tracy–Widom limits and rates of convergence. *Annals of Statistics*, **36**(6), 2638–2716. arXiv:0803.3408.

Karlin, S., and McGregor, J. (1959a). Coincidence probabilities. *Pacific Journal of Mathematics*, **9**(4), 1141–1164.

Karlin, S., and McGregor, J. (1959b). Coincidence properties of birth and death processes. *Pacific Journal of Mathematics*, **9**(4), 1109–1140.

Kasteleyn, P. W. (1961). The statistics of dimers on a lattice: I. The number of dimer arrangements on a quadratic lattice. *Physica*, **27**(12), 1209–1225.

Kasteleyn, P. W. (1963). Dimer statistics and phase transitions. *Journal of Mathematical Physics*, **4**(2), 287–293.

Kasteleyn, P. W. (1967). Graph theory and crystal physics. In F. Harary, ed., *Graph Theory and Theoretical Physics*. Academic Press.

Kenyon, R. (1997). Local statistics of lattice dimers. *Annales de l'Institut Henri Poincaré, Probability and Statistics*, **33**, 591–618. arXiv:math/0105054.

Kenyon, R. (2000). Conformal invariance of domino tiling. *Annals of Probability*, **28**, 759–795. arXiv:math-ph/9910002.

Kenyon, R. (2001). Dominoes and the Gaussian free field. *Annals of Probability*, **29**, 1128–1137. arXiv:math-ph/0002027.

Kenyon, R. (2008). Height fluctuations in the honeycomb dimer model. *Communications in Mathematical Physics*, **281**, 675–709. arXiv:math-ph/0405052.

Kenyon, R. (2009). Lectures on dimers. In *Statistical Mechanics*, IAS/Park City Mathematics Series, vol. 16. American Mathematical Society, 191–230. arXiv:0910.3129.

Kenyon, R. (2014). Conformal invariance of loops in the double-dimer model. *Communications in Mathematical Physics*, **326**, 477–497. arXiv:1105.4158.

Kenyon, R., Lam, W. J., Ramassamy, S., and Russkikh, M. (2018). Dimers and circle packings [Preprint]. arXiv:1810.05616.

Kenyon, R., and Okounkov, A. (2006). Planar dimers and Harnack curves. *Duke Mathematical Journal*, **131**(3), 499–524. arXiv:math/0311062.

Kenyon, R., and Okounkov, A. (2007). Limit shapes and the complex Burgers equation. *Acta Mathematica*, **199**(2), 263–302. arXiv:math-ph/0507007.

Kenyon, R., Okounkov, A., and Sheffield, S. (2006). Dimers and amoebae. *Annals of Mathematics*, **163**, 1019–1056. arXiv:math-ph/0311005.

Kenyon, R., Sun, N., and Wilson, D. B. (2016). On the asymptotics of dimers on tori. *Probability Theory and Related Fields*, **166**, 971–1023. arXiv:1310.2603.

Kerov, S. (2003). *Asymptotic Representation Theory of the Symmetric Group and Its Applications in Analysis*. American Math Society.

Koekoek, R., Lesky, P. A., and Swarttouw, R. F. (2010). *Hypergeometric Orthogonal Polynomials and Their q-Analogues*. Springer Verlag.

König, W. (2005). Orthogonal polynomial ensembles in probability theory. *Probability Surveys*, **2**, 385–447. arXiv:math/0403090.

Korepin, V. E. (1982). Calculation of norms of Bethe wave functions. *Communications in Mathematical Physics*, **86**, 391–418.

Krattenthaler, C. (1990). Generating functions for plane partitions of a given shape. *Manuscripta Mathematica*, **69**(1), 173–201.

Krattenthaler, C. (1999a). Advanced determinant calculus. In D. Foata and G. N. Han, eds., *The Andrews Festschrift*. Springer. arXiv:math/9902004.

Krattenthaler, C. (1999b). Another involution principle-free bijective proof of Stanley's hook-content formula. *Journal of Combinatorial Theory: Series A*, **88**(1), 66–92. arXiv:math/9807068.

Krattenthaler, C. (2002). A (conjectural) 1/3–phenomenon for the number of rhombus tilings of a hexagon which contain a fixed rhombus. In A. K. Agarwal, B. C. Berndt, C. F. Krattenthaler, G. L. Mullen, K. Ramachandra, and M. Waldschmidt, eds., *Number Theory and Discrete Mathematics, Trends in Mathematics*. Birkhäuser. arXiv:math/0101009.

Krattenthaler, C. (2016). Plane partitions in the work of Richard Stanley and his school. In P. Hersh, T. Lam, P. Pylyavskyy, and V. Reiner, eds., *The Mathematical Legacy of Richard P. Stanley*. American Math Society. arXiv:1503.05934.

Kuijlaars, A. B. J. (2011). Universality. In G. Akemann, J. Baik, and P. Di Francesco, eds., *The Oxford Handbook of Random Matrix Theory*. Oxford University Press.

Lai, T. (2017). A q-enumeration of lozenge tilings of a hexagon with three dents. *Advances in Applied Mathematics*, **82**, 23–57. arXiv:1502.05780.

Laslier, B., F. L. and Toninelli, F. L. (2015). Lozenge tilings, Glauber dynamics and macroscopic shape. *Communications in Mathematical Physics*, **338**(3), 1287–1326. arXiv:1310.5844.

Levin, D. A., Peres, Y., and Wilmer, E. L. (2017). *Markov Chains and Mixing Times*, 2nd ed. American Math Society.

Lindström, B. (1973). On the vector representations of induced matroids. *Bulletin of the London Mathematical Society*, **5**(1), 85–90.

Lubinsky, D. S. (2016). An update on local universality limits for correlation functions generated by unitary ensembles. *SIGMA (Symmetry, Integrability and Geometry: Methods and Applications)*, **12**, Article 078. arXiv:1604.03133.

Luby, M., Randall, D., and Sinclair, S. (1995). Markov chain algorithms for planar lattice structures. In *36th Annual Symposium on Foundations of Computer Science*. IEEE COmputer Society.

Macdonald, I. G. (1995). *Symmetric Functions and Hall Polynomials*, 2nd ed. Oxford University Press.

MacMahon, P. A. (1896). Memoir on the theory of the partition of numbers. Part I. *Philosophical Transactions of the Royal Society of London, Series A*, **187**, 619–673.

MacMahon, P. A. (1960). *Combinatory Analysis*, vol. 1–2. Chelsea.

Matytsin, A. (1994). On the large N limit of the Itzykson–Zuber integral. *Nuclear Physics B*, **411**(2–3), 805–820. arXiv:hep-th/9306077.

McCoy, B. M., and Wu, T. T. (1973). *The Two-Dimensional Ising Model*. Harvard University Press.

Mehta, M. L. (2004). *Random Matrices*, 3rd ed. Elsevier/Academic Press.

Metcalfe, A. (2013). Universality properties of Gelfand–Tsetlin patterns. *Probability Theory and Related Fields*, **155**(1–2), 303–346. arXiv:1105.1272.

Mkrtchyan, S. (2019). Turning point processes in plane partitions with periodic weights of arbitrary period [Preprint]. arXiv:1908.01246.

Montroll, E. W., Potts, R. B., and Ward, J. C. (1963). Correlations and spontaneous magnetization of the two-dimensional Ising model. *Journal of Mathematical Physics*, **4**, 308–322.

Morales, A. H., Pak, I., and Panova, G. (2019). Hook formulas for skew shapes III. Multivariate and product formulas. *Algebraic Combinatorics*, **2**(5), 815–861. arXiv:1707.00931.

Nekrasov, N. (2016). BPS/CFT correspondence: non-perturbative Dyson–Schwinger equations and qq-characters. *Journal of High Energy Physics*, **2016**(181), arXiv:1512.05388.

Neretin, Y. A. (2003). Rayleigh triangles and non-matrix interpolation of matrix beta integrals. *Sbornik: Mathematics*, **194**(4), 515–540. arXiv:math/0301070.

Nica, A., and Speicher, R. (2006). *Lectures on the Combinatorics of Free Probability*. Cambridge University Press.

Nienhuis, B., Hilhorst, H. J., and Blöte, H. W. J. (1984). Triangular SOS models and cubic-crystal shapes. *Journal of Physics A: Mathematical and General*, **17**, 3559–2581.

Nordenstam, E. (2009). Interlaced particles in tilings and random matrices. Unpublished Ph.D. thesis, KTH Royal Institute of Technology.

Nordenstam, E., and Young, B. (2011). Domino shuffling on Novak half-hexagons and Aztec half-diamonds. *Electronic Journal of Combinatorics*, **18**, article P181. arXiv:1103.5054.

Okounkov, A., and Reshetikhin, N. (2003). Correlation function of Schur processes with application to local geometry of a random 3-dimensional Young diagram. *Journal of the American Mathematical Society*, **16**(3), 581–603. arXiv:math/0107056.

Okounkov, A., and Reshetikhin, N. (2006). The birth of a random matrix. *Moscow Mathematical Journal*, **6**(3), 553–566.

Okounkov, A., and Reshetikhin, N. (2007). Random skew plane partitions and the Pearcey process. *Communications in Mathematical Physics*, **269**(3), 571–609. arXiv:math/0503508v2.

Olshanski, G., and Vershik, A. (1996). Ergodic unitarily invariant measures on the space of infinite Hermitian matrices. In R. L. Dobrushin, R. A. Minlos, M. A. Shubin, and A. M. Vershik, eds., *Contemporary Mathematical Physics. F. A. Berezin's Memorial Volume*, series 2, vol. 17. American Mathematical Society. arXiv:math/9601215.

Pak, I., Sheffer, A., and Tassy, M. (2016). Fast domino tileability. *Discrete & Computational Geometry*, **56**(2), 377–394. arXiv:1507.00770.

Petrov, L. (2014a). Asymptotics of random lozenge tilings via Gelfand–Tsetlin schemes. *Probability Theory and Related Fields*, **160**(3–4), 429–487. arXiv:1202.3901.

Petrov, L. (2014b). The boundary of the Gelfand–Tsetlin graph: new proof of Borodin–Olshanski's formula, and its q-Analogue. *Moscow Mathematical Journal*, **14**, 121–160. arXiv:1208.3443.

Petrov, L. (2015). Asymptotics of uniformly random lozenge tilings of polygons. Gaussian free field. *Annals of Probability*, **43**(1), 1–43. arXiv:1206.5123.

Pickrell, D. (1991). Mackey analysis of infinite classical motion groups. *Pacific Journal of Mathematics*, **150**, 139–166.

Prähofer, M., and Spohn, H. (2002). Scale invariance of the PNG droplet and the Airy process. *Journal of Statistical Physics*, **108**(5–6), 1071–1106. arXiv:math/0105240.

Prasolov, V. (1994). *Problems and Theorems in Linear Algebra*. Vol. 134 of *Translations of Mathematical Monographs*. American Mathematical Society.

Propp, J. (2003). Generalized domino-shuffling. *Theoretical Computer Science*, **303**, 267–301. arXiv:math/0111034.

Propp, J. (2015). Enumeration of Tilings. In Miklos Bona, ed., *Handbook of Enumerative Combinatorics (Discrete Mathematics and Its Applications)*. CRC Press.

Propp, J., and Wilson, D. (1996). Exact sampling with coupled Markov chains and applications to statistical mechanics. *Random Structures and Algorithms*, **9**(1–2), 223–252.

Purbhoo, K. (2008). Puzzles, tableaux, and mosaics. *Journal of Algebraic Combinatorics*, **28**(4), 461–480, arXiv:0705.1184.

Quastel, J., and Spohn, H. (2015). The one-dimensional KPZ equation and its universality class. *Journal of Statistical Physics*, **160**(4), 965–984. arXiv:1503.06185.

Randall, D., and Tetali, P. (2000). Analyzing Glauber dynamics by comparison of Markov chains. *Journal of Mathematical Physics*, **41**, 1598.

Romik, D. (2015). *The Surprising Mathematics of Longest Increasing Subsequences*. Cambridge University Press.

Rosengren, H. (2016). Selberg integrals, Askey–Wilson polynomials and lozenge tilings of a hexagon with a triangular hole. *Journal of Combinatorial Theory, Series A*, **138**, 29–59 1503.00971

Russkikh, M. (2018). Dimers in piecewise Temperleyan domains. *Communications in Mathematical Physics*, **359**(1), 189–222. arXiv:1611.07884.

Russkikh, M. (2019). Dominoes in hedgehog domains. *Annales de l'Institut Henir Poincare D*, **8**(1), 1–33. arXiv:1803.10012.

Sagan, B. E. (2001). *The Symmetric Group: Representations, Combinatorial Algorithms, and Symmetric Functions*, 2nd ed. Springer.

Sato, R. (2019). Inductive limits of compact quantum groups and their unitary representations [Preprint]. arXiv:1908.03988.

Sheffield, S. (2005). Random surfaces. *Astérisque*, **304**, 181. arXiv:math/0304049.

Sheffield, S. (2007). Gaussian free fields for mathematicians. *Probability Theory and Related Fields*, **139**, 521–541. arXiv:math/0312099.

Shiryaev, A. N. (2016). *Probability 1*, 3rd ed. Vol. 95 of Graduate Texts in Mathematics. Springer-Verlag.

Shlosman, S. (2001). Wulff construction in statistical mechanics and in combinatorics. *Russian Mathematical Surveys*, **56**(4), 709–738. arXiv:math-ph/0010039.

Tao, T., and Vu, V. (2012). Random matrices: the universality phenomenon for Wigner ensembles. In *Proceedings of Symposia in Applied Mathematics*, **2014**(72). arXiv:1202.0068.

Temperley, H. N. V., and Fisher, M. E. (1961). Dimer problem in statistical mechanics—an exact result. *Philosophical Magazine*, **6**(68), 1061–1063.

Thiant, N. (2003). An O (n log n)-algorithm for finding a domino tiling of a plane picture whose number of holes is bounded. *Theoretical Computer Science*, **303**(2–3), 353–374.

Thurston, W. P. (1990). Conway's tiling groups. *The American Mathematical Monthly*, **97**(8), 757–773.

Toninelli, F. L. (2019). Lecture notes on the dimer model. `http://math.univ-lyon1.fr/homes-www/toninelli/noteDimeri_latest.pdf`.

Vershik, A. (1997). Talk at the 1997 conference on Formal Power Series and Algebraic Combinatorics, Vienna.

Vershik, A. M., and Kerov, S. V. (1982). Characters and factor representations of the infinite unitary group. *Soviet Mathematics Doklady*, **26**, 570–574.

Voiculescu, D. (1976). Représentations factorielles de type II_1 de $U(\infty)$. *Journal de Mathématiques Pures et Appliquées*, **55**, 1–20.

Voiculescu, D. (1991). Limit laws for random matrices and free products. *Inventiones Mathematicae*, **104**, 201–220.

Voiculescu, D., Dykema, K., and Nica, A. (1992). *Free random variables*. Vol. 1 of CRM Monograph Series 1. American Mathematical Society.

Werner, W., and Powell, E. (2020). Lecture notes on the Gaussian Free Field [Preprint]. arXiv:2004.04720.

Weyl, H. (1939). *The Classical Groups: Their Invariants and Representations*. Princeton University Press.

Wilson, D. (1997). Determinant algorithms for random planar structures. In M. Saks, ed., *Proceedings of the Eighth Annual ACM-SIAM Symposium on Discrete Algorithms*. Society for Industrial and Applied Mathematics.

Wilson, D. (2004). Mixing times of Lozenge tiling and card shuffling Markov chains. *Annals of Applied Probability*, **14**(1), 274–325. arXiv:math/0102193.

Young, B. (2010). Generating functions for colored 3D Young diagrams and the Donaldson–Thomas invariants of orbifolds (with appendix by Jim Bryan). *Duke Mathematical Journal*, **152**(1), 115–153.

Zeilberger, D. (1996). Reverend Charles to the aid of Major Percy and Fields-Medalist Enrico. *The American Mathematical Monthly*, **103**(6), 501–502. arXiv:math/9507220.

Zinn-Justin, P. (2009). Littlewood–Richardson coefficients and integrable tilings. *Electronic Journal of Combinatorics*, **16**, article R12. arXiv:0809.2392.

Index

Printed in the United States
by Baker & Taylor Publisher Services